D0984971

CONSTRUCTION ELECTRICAL CONTRACTING

Wiley Series of Practical Construction Guides

M. D. Morris, P. E., Editor

Jacob Feld
CONSTRUCTION FAILURE

William G. Rapp
CONSTRUCTION OF STRUCTURAL STEEL BUILDING FRAMES

John Philip Cook
CONSTRUCTION SEALANTS AND ADHESIVES

Ben C. Gerwick, Jr.
CONSTRUCTION OF PRESTRESSED CONCRETE STRUCTURES

S. Peter Volpe
CONSTRUCTION MANAGEMENT PRACTICE

Robert Crimmins, Reuben Samuels, and Bernard Monahan
CONSTRUCTION ROCK WORK GUIDE

B. Austin Barry
CONSTRUCTION MEASUREMENTS

D. A. Day
CONSTRUCTION EQUIPMENT GUIDE

Harold J. Rosen
CONSTRUCTION SPECIFICATION WRITING

Gordon A. Fletcher and Vernon A. Smoots
CONSTRUCTION GUIDE FOR SOILS AND FOUNDATIONS

Don A. Halperin
CONSTRUCTION FUNDING: WHERE THE MONEY COMES FROM

Walter Podolny, Jr., and John B. Scalzi
CONSTRUCTION AND DESIGN OF CABLE-STAYED BRIDGES

John P. Cook
COMPOSITE CONSTRUCTION METHODS

William B. Kays
CONSTRUCTION OF LININGS FOR RESERVOIRS, TANKS, AND POLLUTION CONTROL FACILITIES

John E. Traister
CONSTRUCTION ELECTRICAL CONTRACTING

CONSTRUCTION ELECTRICAL CONTRACTING

JOHN E. TRAISTER

John E. Traister Associates

A Wiley-Interscience Publication

JOHN WILEY & SONS

New York • Chichester • Brisbane • Toronto • Singapore

Library of Congress Cataloging in Publication Data

Traister, John E.
Construction electrical contracting.

(Wiley series of practical construction guides)
"A Wiley-Interscience publication."
Includes index.
1. Electric contracting—United States. I. Title.

HD9695.U52T73 658'.92'13 78-13441
ISBN 0-471-02986-6

Printed in the United States of America

10 9 8 7 6 5 4

Series Preface

The construction industry in the United States and other advanced nations continues to grow at a phenomenal rate. In the United States alone construction in the near future will exceed ninety billion dollars a year. With the population explosion and continued demand for new building of all kinds, the need will be for more professional practitioners.

In the past, before science and technology seriously affected the concepts, approaches, methods, and financing of structures, most practitioners developed their know-how by direct experience in the field. Now that the construction industry has become more complex there is a clear need for a more professional approach to new tools for learning and practice.

This series is intended to provide the construction practitioner with up-to-date guides which cover theory, design, and practice to help him approach his problems with more confidence. These books should be useful to all people working in construction: engineers, architects, specification experts, materials and equipment manufacturers, project superintendents, and all who contribute to the construction or engineering firm's success.

Although these books will offer a fuller explanation of the practical problems which face the construction industry, they will also serve the professional educator and student.

M.D. Morris, P.E.

Preface

Each year a number of individuals enter the electrical contracting business without giving too much thought to some of the problems, strengths, and weaknesses particularly evident in this highly competitive business. Lack of knowledge of these problems, or failure to consider them, usually dooms the business operation to failure within two or three years.

Construction Electrical Contracting is designed to orient the reader—whether he is thinking of entering the business or is already active in the industry—to the responsibilities and risks involved in the business of electrical contracting, the technical performance required, and all other aspects of the business. A knowledge of all phases covered in this book will help to provide a more profitable business operation and to prevent failure.

For employees, the contents of this book should further establish an awareness of the factors involved in the field so that they will have a better understanding of the problems confronting their employers. With this understanding, they should become more responsible workers with a sincere interest in the success of the firm's operation.

JOHN E. TRAISTER

Bentonville, Virginia
June 1978

Contents

CONSTRUCTION ELECTRICAL CONTRACTING

1

Introduction to Electrical Contracting

This chapter is directed mainly towards the prospective electrical contractor. However, the problems of sufficient cash and credit, office equipment, tools and equipment, office location, layout, and costs, as well as obtaining qualified personnel would no doubt be of great interest to the existing contractor also.

1.1 Prerequisite of an Electrical Contractor

Electrical contracting can be a profitable business, but before entering into such a venture, one must be familiar with all the essentials for successful operation. There are many hazards of getting started—many more than most people realize—and the sooner the potential electrical contractor realizes this, the sooner he will be able to operate the business with reasonable smoothness and impunity.

A person who is considering entering the electrical contracting business must first take a long, hard look at himself to determine if he is in fact qualified to handle such a business. If he is not satisfying his current employer, chances are he won't satisfy his future customers in the contracting business. On the other hand, a well-trained electrician who has proven that he has a good knowledge of the National Electrical Code, is familiar with modern electrical installation techniques, and is able to direct men well, at least has one foot in the door.

Still, just because a person may be proficient in electrical construction work doesn't mean that he is guaranteed success in the electrical construction business. Too often, people emphasize their strong points and fail to recognize their weak ones. For example, an electrician may be an excellent mechanic, foreman, superintendent, or estimator but, owing to lack of sufficient financing, executive ability, or general knowledge of the business, he is not qualified to operate an electrical contracting firm.

The requirements of a person entering the business are many but, aside from a knowledge of electrical installation techniques, confidence and enthusiasm are probably two of the most important traits a potential electrical contractor can have. With confidence in himself and enthusiasm to conquer obstacles and make the business run on a profitable basis, most level-headed, well-trained electricians will have a chance.

The prospective electrical contractor should further have a good knowledge of the National Electrical Code, a relatively good knowledge of electrical design, reasonable mathematical ability, the ability to accomplish goals through men (not by men), the willingness to make sacrifices, a good knowledge of electrical estimating, and a good knowledge of business fundamentals. Any prospective electrical contractor who lacks any of the above-mentioned requirements should take steps to acquire them.

To illustrate, if a person is weak in interpreting the NEC, there are many night courses available throughout the country on this subject; check with your local building inspector for details. Other courses are offered by junior colleges, vocational, and other schools on all phases of electrical construction. There are also many good books on the subjects suitable for self-study. International Correspondence School (Scranton, Pennsylvania) offers a course titled *Electrical Contracting* which is ideal for the prospective electrical contracting. Colleges are now offering degree programs in electrical contracting. The sources are available; one just has to take time to get them.

1.2 Proper Direction of Personnel

Since electrical contracting entails getting an electrical installation completed in the shortest possible time, in a workmanlike manner, and with the least possible expense, it follows that the man leading such an organization must be a rather highly qualified person. What are some of the qualifications of an electrical contractor in regard to directing personnel? First, the electrical contractor must be a leader. Not one who merely gives orders, but one capable of stimulating job interest and have his men turn out a good job not necessarily because they have to, but because they want to.

Ideally, the electrical contractor should have the ability to do almost any electrical job as well as, if not better than, the men under his supervision. He should further be an organizer with the ability to organize his men, equipment, and materials in an orderly manner that will produce maximum efficiency. His judgment should inspire confidence and his character should be above reproach.

There are also other factors like education, physical and emotional stability, interest in the morale and welfare of his men, which encourage the men to respect the contractor, enabling him to properly direct them better.

The electrical contractor should further strive to attain the following characteristics:

1. Ability to:
   a. Exercise good judgment
   b. Be realistic
   c. Make good and prompt decisions
   d. Refrain from "passing the buck"
   e. Establish good communications with others
   f. Assign workers to the best advantage
   g. Induce respect and loyalty from others.
2. Know his own capabilities, faults, and be completely objective.
3. Exhibit enthusiasm and a dynamic drive for accomplishment.
4. Be cooperative and induce cooperation from others.
5. Be industrious and induce industriousness in others.
6. Be patient, calm, and even-tempered.
7. Be friendly.
8. Exhibit good management manners.
9. Recognize workers as human beings.
10. Be sensitive to human traits.
11. Have a good character.

Once an electrical contractor has gained all the preceding characteristics, he cannot help but have the ability to properly direct his personnel.

1.3 Knowledge of Estimating Procedures

An estimate of an electrical construction installation is the basis of providing the types and quantities of materials needed for completion of the project; the required man-hours; extent of premium time anticipated; the number of workers anticipated to be required at various intervals; and similar items.

All the factors in the previous paragraph must be taken into consideration by the estimator in preparing the estimate. Other factors—in some cases—may be necessary.

In general, electrical estimating is the determination of a sound cost and proper price for a given amount of work in advance of performing the work. While no estimate can be absolutely correct (except by accident), most can be surprisingly close to the actual cost of the project.

The procedures necessary in making an estimate of an electrical project consist of making an accurate count of all outlets, lighting fixtures, conduit, cable, and other items of material and then listing them on a pricing sheet. Labor and direct job expenses are then applied to the material items to arrive at an estimated price cost of the project. To this, overhead and profit are added to arrive at a bid price.

Those who perform electrical estimates must also be aware of the many job factors that will vary from job to job. The estimator must also take into consideration the depreciation and consumption of tools, job supervision, purchasing expenses, interest on borrowed money, and so on.

A detailed explanation of electrical estimating procedures is covered in later chapters. However, the prospective electrical contractor should study several textbooks written solely on this important phase of the electrical contractor's business in order to gain a thorough knowledge of the various methods and techniques. Several courses on electrical estimating are also available which should be of great aid to both the prospective and veteran contractor. International Correspondence School offers several lessons in electrical estimating in its course titled *Electrical Contracting* and the National Electrical Contractor's Association (NECA) has estimating courses available for members of the organization.

There are also several labor unit manuals for sale or lease on the market and most of these give detailed explanations of the manual's use in estimating electrical construction. It is recommended that the prospective electrical contractor investigate all of these estimating methods and then select the one system that seems to suit his individual needs the best.

1.4 Sufficient Cash and Credit

Many electrical contractors—especially those starting out on residential construction—have been able to get by on less than $1,000 cash. However, for a business of any size, $10,000 to $25,000 cash should be considered absolute minimum. There are many who have tried starting a business to do large construction work on less than this amount, using credit almost entirely, but most of them are still trying to get started.

The exact cost of starting an electrical contracting business depends on several factors, but the two most important considerations are the type of business and the volume of work anticipated. A one-man business, for example, can begin on a relatively small amount of money if the owner does all the estimating, design, selling, and administrative work. Then if the owner is able to hire four or five good electricians—those capable of installing electrical systems with few instructions—the owner may be able to survive for a while on say, $7500 initial cash investment, if he is willing to put on from 60 to 70 hours per week.

On the other hand, if the contractor plans to equip an office and hire an estimator and bookkeeper, and hire from 8 to 20 electricians, $25,000 initial cash is none too much.

Before anyone starts an electrical contracting business, he is certainly going to have some work already lined up. Even if he does, he will still have to wait at least one month (and often two months) before the first checks begin to arrive. If you figure up the expenses for one month on a conservative basis, here is what the results would be.

Owner's salary	$1,200.00
Secretary	400.00
Rent (including utilities)	600.00
Interest on investment	120.00
Automobile and travel expenses	255.00
Insurance	60.00
Wages for four workers	7,120.00
Payment of office equipment, tools, etc.	400.00
Miscellaneous	600.00
	$10,755.00

Now if your customer was, say, 30 days late in getting the check to you, this would mean that your expenses would be $21,510 before you received your first check. You might think that your creditors may wait until you get the check before you will have to pay the bills. Unfortunately, this is rarely the case. Your employees will want their paychecks weekly (they have bills to pay also); the utility companies must be paid promptly or they will stop service; and so on.

Notice that we didn't include any payment for materials. Normally, an electrical supplier will extend 60- to 90-day credit for the first period a new contractor is in business to help him get started, provided prior arrangements are made. However, if a supplier will not extend such credit, then you will need from $10,000 to $20,000 additional funds to pay for materials. It should now be obvious why an electrical contractor needs sufficient working capital to even start a business.

Besides the cash mentioned previously, the electrical contractor should also have a line of credit equal to three months' operating expenses. Therefore, if the total estimated monthly operating expenses is $25,000 the first year, the electrical contractor should have a line of credit equal to about $75,000. This, of course, is going to take security, and often means mortgaging one's home and other assets to obtain such credit.

In any event, the prospective contractor should investigate the sources of his anticipated finances, and then make certain that the funds or line of credit will be available at the time he needs them. A letter of confirmation from a bank and electrical supplier should be obtained.

1.5 Office Equipment

The type and amount of office equipment will naturally vary depending upon the type and volume of work anticipated by the contractor. For a one-man operation, however, the following should be considered minimum.

1	– Secretary's desk/chair	$350.00
1	– Owner's desk/chair	400.00
1	– Estimating table/stool	125.00
1	– Four-drawer file cabinet	165.00
1	– Plan rack	40.00

1 –	Typewriter	500.00
1 –	Safe (optional)	500.00
1 –	Bookcase (reference books and catalogs)	60.00
1 –	Drafting table	300.00
2 –	Chairs for customers	100.00
		$2,540.00

This is rather skimpy, but a new contractor (one-man operation) can get by with just these items until profits warrant buying more equipment. Of course, the contractor will need other miscellaneous items such as letterhead, envelopes, estimating and drafting tools, bookkeeping supplies, an electronic calculator, and so on, but these items will normally not cost more than $500 to $600.

From the preceding, we can see that a new one-man contracting business will require at least $3000 to equip an office, more if additional employees are hired.

1.6 Tools and Equipment

The type and amount of tools that a contractor will need to begin operation again depends upon the type and volume of business. A one-man operation doing small commercial and residential projects, for example, will normally have little more than a workbench, bench-mounted pipe vise, machinist's vise, drill stand, and perhaps an electric grinder in the shop. Any other tools required for work in the shop is done with portable tools (drill motors, pipe benders, etc.) that may also be used on jobs in the field.

Any contracting firm, regardless of the size, will normally have at least one truck and an assortment of tools. The truck should be equipped with the following tools:

1 – ¼″ Electric drill
1 – ½″ Electric drill
  Bit extensions
  Miscellaneous wood bits
1 – Pipe vise and stand
1 – Machinist vise mounted on truck
1 – Combination ½″, ¾″, and 1″ stock and dies
1 – Hand ratchet 1¼″–2″ stock and dies
1 – ½″ Conduit hickey
1 – ¾″ Conduit hickey
1 – 1″ Conduit hickey
1 – ½″ EMT bender
1 – ¾″ EMT bender
1 – Fish tape
1 – 6′ Step ladder
1 – 10′ Step ladder
1 – 20′ Extension ladder

Larger projects will require hydraulic pipe benders, several drill motors, cable cutters, power threading dies, and the like, just to mention a few.

In general, the smallest contractor should count on at least $3000 initial investment for tools, and then a regular allowance to replace expendable items such as drill bits, hack saw, blades, and so on.

Tools and equipment can definitely be time-saving devices that save the contractor much time and money, but one must never overrate them. A contractor who invests heavily in a tool inventory can run into problems. For example, assume that a contractor has $20,000 worth of tools and equipment. Should there be a slack in the contractor's work load, or if his work stops altogether, who is going to have any use for the tools? The contractor certainly won't, and chances are neither will other contractors in the area, they might also be out of work! So the tools can't be used and can't be sold at a reasonable price. All the contractor has is a lot of money tied up in a tool inventory that is worthless at the moment.

1.7 Office Location and Layout

The physical arrangement of the shop and office is an important factor in providing for the efficient servicing of the work with materials and tools at the lowest possible cost.

In some instances, because of the limitations of type of building, floor area, type of building access, relation to streets, yard area, and so on, it may not always be possible to provide an ideal physical arrangement of facilities; however, intelligent planning can usually provide for the maximum efficiency under the given conditions.

The primary objectives of a good physical arrangement will be to provide for: (1) ease of handling materials and tools with a minimum of shifting around and extra handling; (2) ready accessibility; (3) sufficient predelivery storage or holding areas for assembled orders for specific jobs and holding areas for materials and tools returned from jobs, awaiting checking and return to inventory; (4) unobstructed truck access; (5) centrally located stock room desk area or office.

In general these objectives are best accomplished when: (1) all facilities are located on one floor; (2) adequate floor area is available; (3) there are sufficient bins, shelving, and storage racks with adequate aisles and walkspace; (4) separate receiving–unloading and shipping–loading area or driveways can be provided; (5) truck-bed height platforms for unloading, storage, and loading of cable reels and heavy equipment are provided; (6) yard space accessible to trucks is available for parking, storage, receiving, and shipping, preferably with receiving and shipping entrances opening from the building to the yard space; (7) driveway access from the street to the building or yard space can be made from streets that are not congested with traffic.

The entire arrangement should be such that production line sequence of handling can be approached to as great an extent as possible. Related materials that are normally shipped at the same time should be grouped in reasonable proximity in line with the frequency of their shipment.

When available floor area is restricted, balconies can often be constructed which will increase the total floor area and which can be used for the storage of items which are handled and shipped less frequently.

1.8 Office Personnel

As any electrical contractor knows, the top-level supervisory personnel are very important people to the firm. They must be executives, administrators, salesmen, often politicians, and have a good working knowledge of the various codes and ordinances, along with a fair knowledge of building construction in general.

The electrical estimator in particular is one of the key men in any electrical contractor's operation and must bear a heavy load of responsibility. Basically, this responsibility includes the ability to determine—through accurate quantity surveys and intelligent analysis of all variable conditions—a reasonably accurate cost and price that will provide a sufficient amount of income from a given job to defray all the costs of material, labor, direct job expenses caused by the job, and a proper share of the overhead or operating expenses, and leave a normal margin of profit for the firm. Any estimator who does not accomplish this can break a firm in a very short period of time.

Consequently, such a man is a rare bird indeed, as most contractors have found to their sorrow. He is hard to find, and once found and geared to the contractor's operation, he is hard to keep from the clutches of other predatory contractors.

1.8.1 Potential Estimators

Where does an electrical contractor look for electrical estimators? The current trend seems to be the hiring of graduate engineers fresh out of college before the larger industries can grab them. If such a person is able to keep his feet on the ground and learn the practical applications of his college training—perhaps as a helper for an electrical contractor during college vacations—he should have the capabilities of making a good electrical estimator.

Although college graduates are, by no means, the only source of potential electrical estimators, any candidate for the position should possess a few personal characteristics necessary in a successful estimator. These characteristics include:

1. *Reasonable mathematical ability:* Usually two years of high school algebra and one year of plane geometry is more than sufficient.
2. *An orderly mind* is essential in the profession of electrical estimating.
3. *Experience:* The majority of electrical estimators have had some experience in the electrical construction industry working on actual construction projects.
4. He is *personable* because he has to deal with prospective customers, consulting engineers, salesmen, construction superintendents and other persons in the construction industry at large.
5. He is usually *married*, and probably has one or more children. The reason? A man with responsibilities will tend to stick with one contractor longer, rather than gamble on a greener-grass offer from another contractor.
6. He has *imagination and ability to visualize* the drawings as a completed project or any phase required for the completion.
7. He has the *ability to concentrate* on his work without letting personal or other problems interfere.
8. He has the *patience* to be very thorough in his work.

1.8.2 Where to Find the Estimator

Advertising in a local newspaper or trade journal probably accounts for most of the job applicants. Still, the potential estimator may be right under your nose. Perhaps he is even working for you right now—in a different position.

How about the apprentice electrician who is working on one of your jobs? He may have dropped out of college after a year or two and is now working towards a different goal. But he could be just the man for training as a mechanical estimator—especially at his wages as compared to that of a graduate engineer's salary.

A salesman working for an electric equipment supply company is another possibility. Such people have usually had extensive experience in quantity take-off for their customers. An electrical draftsman working in a consulting engineers office could perhaps be another possibility. He may have reached a "dead-end" situation in the consulting firm because of his lack of college education, but if he is well trained, he could make a fine estimator.

Yes, the possibilities are great, but getting the word to these sources may not be easy. Many of the potential estimators working in other fields may not be aware that they are qualified for the job. The electrical contractor will have to use his imagination if he desires to get the word across.

An electrical estimator, for example, is probably going to look under "Draftsman" in the classified sections of newspapers and magazines and not under "Estimator." So the contractor may advertise as follows:

Electrical draftsman with four or more years experience wanted as an estimator trainee for local electrical contracting firm.

A notice posted on your job site bulletin board may produce a latent electrical estimator who is now your job foreman.

1.8.3 How to Keep Employees

Money, needless to say, is one of the best methods of keeping employees happy. Pay the estimator what he is worth and make certain that he receives pay increases immediately when they are warranted. There have been several cases when a delay of two weeks on a pay increase has cost the contractor a good estimator; he went elsewhere, where the contractor was a little more prompt.

The profit-sharing plan is also a help, but making the employees wait five or ten years to get their hands on the money is often dangerous. Profit-sharing in the form of a yearly bonus often works best.

The type of treatment the employee receives is also of great importance. A compliment for a job well done is worth more, in some cases, than a pay increase. Also make certain that the employee is complimented in front of other employees (and chewed out in private); this not only makes the person feel good, but it may tend to promote a higher productivity in the other departments, as they will be trying for a similar compliment.

1.9 Obtaining Qualified Electrical Workers

Those electrical contractors who choose to be affiliated with the I.B.E.W. or other electrical union will normally have a pool of qualified electrical workers to choose from at any given time. However, in times of large construction projects in the area, the union may not be able to fill all the needs of all contractors. If the prospective electrical contractor is not familiar with the policies of organized labor, it is suggested that he meet with the business agent of the jurisdiction in which the contractor is located and discuss the situation.

Open shop contractors usually advertise in local newspapers for electricians and helpers or contact friends in the electrical construction industry who may be willing to change jobs or else can recommend other electricians who might.

The prospective contractor as well as the veteran should stay aware what is happening with other firms in the area. Perhaps another contractor is going out of business or is having a lay-off due to a slack in the firm's work. This is a good time to pick up good electricians.

Many open shop contractors take on a few helpers and train them on the job with the hopes that they will remain with the firm for a long period of time after they learn the trade. However, this is not guaranteed, and many times several of these employees will look for greener pastures after they become qualified electricians.

2

Preliminary Construction Techniques

This chapter outlines practical and efficient preliminary construction techniques which will serve as a guide to the electrical contractor and his/her personnel. These procedures are considered necessary, in conducting this phase of the business operation, to lay a foundation for reducing the direct and indirect costs of performing and servicing the work to the lowest possible minimum, thus helping to assure a profitable business operation.

2.1 Obtaining Permits and Notices

Once the contract has been awarded, the electrical contractor should not delay in applying for any required building permits since work cannot be started until they are received. Some contractors have waited until the last minute to obtain a permit and then find that the project is delayed because the building department requires an inconveniently long time to review the plans and specifications.

If this is your first project of a given size in a particular jurisdiction, you should be especially prompt in obtaining the permits, because special qualifications (insurance, bonds, personal qualifications, etc.) may be required. If your firm lacks any of the qualifications, checking into them early will give you sufficient time to qualify without delaying the starting of the project.

In some areas, permits will only be required of the owner and general contractor, relieving the electrical contractor of all responsibilities pertaining to permits with perhaps the exception of an electrical inspection at various intervals during the construc-

Figure 2.1

tion. In any case, check with the local building department in the area where the project is located and if permits are required, they are usually required to be posted or available on the job. Again, the building department can give you all the details.

2.2 Preliminary Conferences

Shortly after the general contract is awarded, the general contractor usually calls a meeting with all of the successful subcontractors. At this meeting, the architect's/engineer's drawing and specifications (including all addendums) are discussed and schedules and manpower requirements are agreed upon. Other subjects of discussion include delivery and storage of materials, cleanup responsibilities, temporary power requirements, location of office and tool storage, workmen's facilities, changes, and claims for extras. This would also be a good time to discuss the dates you will receive progress payments.

Since all of the subcontractors are together at this first meeting, the electrical contractor should discuss the coordination of the various trades at this time—both from the standpoint of the portions of the electrical system placed on or connected to it and the effect of the physical arrangement of those systems on the location and arrangement of the electrical system. Subs that cooperate at this first meeting can also possibly save each other some expenses. An example would be trenching for plumbing and underground electrical raceways. Perhaps the plumbing and electrical contractor can work out an agreement whereas both trades could use a common trench for their pipe runs and then share the expense.

In the case of conflicts—like control wiring for HVAC equipment or limited space in a mechanical room where mechanical and electrical equipment are in each other's way—solutions can often be worked out at these preliminary meetings, saving precious time on the job where dozens of workers would be made to wait while you try to resolve the problem.

Next will be a meeting with the architect, owners, engineers, general contractor, and all subcontractors. The architect will usually take charge at these meetings and inform all concerned just what is expected of them; how bills, shop drawings, and so on are to be submitted. This is also the time to clarify any doubtful aspects of the project. This is *not* the time, however, to point out errors and/or omissions in his drawings; they should be discussed either before or after this meeting. Everyone makes a mistake occasionally, and the architect and engineer are no exceptions. However, no one wants to be embarrassed in front of a crowd of people.

When obvious errors are detected in the architect's or engineer's drawings, such errors should immediately be called to their attention, either directly or through the proper channels so that appropriate action may be taken at the earliest possible time. This should be done diplomatically, as mentioned previously, in order to cause the least embarrassment and to maintain a good relationship with the architect/engineer; the next mistake made on the project could be made by you or your men!

Where little or no extra cost to the electrical contractor is involved, it may be in the best interest of all concerned to install the work correctly without saying too much about it, but if you do, make certain you're right. If you correct something, the final installation will obviously vary from the original drawings, and you're then responsible for making the change.

A close relationship with the architect and engineer cannot be overemphasized. Their purpose is to provide the owner with a sound and useful design and then make certain the contractors carry out their work as called for in the plans and specifications.

Many architects even look to the reliable contractors for a final check on their electrical drawings. For this reason, maintain your good standing with architects and engineers by helping to provide a complete installation on schedule, in a workmanlike manner, and without too many change orders. If you can do this, you will benefit greatly on future jobs. However, the place to begin this relationship is at the first meeting.

2.3 The Basis of Budgeting Performance

In general, the basis of budgeting performance on a given project is simply the anticipation of what will be required of the contractor in the future and preparing to carry out the requirements in the most economical way.

Since it is usually best to start this planning prior to starting actual installations on the job, we would prefer to this portion of the scheduling as "pre-planning and scheduling." In its simplest form, it means that a time and date is selected to start the project; a job foreman has been selected to run the job; arrangements have been made to discuss the project with the selected foreman and to present him with work orders and construction documents. Finally, arrangements will be made for materials and tools to be delivered as well as the necessary workmen.

For the smaller projects, this pre-planning and scheduling may consist of only a mental analysis of the situation on the part of the contractor. The job foreman will then make further decisions when he reports to the project. The foreman should, however, be backed up with good material and tool support for proper management.

The larger projects will normally require considerably more formal planning and scheduling if the job is to be performed in the most economical manner; that is, to perform the installation with the lowest possible man-hours consistent with the job requirements and good workmanship.

Degrees of planning and scheduling can range from notations on the drawings or estimate forms to the use of computers for a more orderly, convenient, and faster procedure. The format can also vary from printouts in tabulations to various types of graphical charts or a combination of both.

In order to carry out good planning and scheduling procedures, the supervisory personnel must have the drive and desire to carry them out. Supervisory personnel should further have qualities of leadership to inspire those under their direction to work efficiently both individually and cooperatively with others. When efficiency is not up to par, they should have instructional ability to acquaint those under their direction with new or improved methods of making various installations and to guide those lacking experience in certain phases of the work to improve the situation.

Supervisory personnel should then have the necessary instructions and information in order to know exactly what is required of them. Basically, instructions and information will be in the form of a work order, plans and specifications, detailed shop drawings, and perhaps some notes from the estimator.

Finally, the job foreman must be provided with the necessary materials, tools, and similar facilites in order to perform the work. This is where planning and scheduling can really pay off. Requisitions for out-of-stock materials should first be written up along with the approximate dates the various items of material will be needed on the job. Purchase orders should be written for the panelboards, lighting fixtures, and similar items that require special order. Again, the delivery date should be determined and followed up later to insure that they are being shipped as scheduled.

A list of tools—enough to begin the job—should be ordered immediately, followed by a complete list for various phases of the construction. For example, it may be determined that a hydraulic 4-in. pipe bender will be needed the third week on the job to install some service conduit in the deck prior to pouring the footing. This gives the contractor time to arrange for such a bender to be sent on the job. It will also give other project foremen a chance to schedule their work so that they can do without the bender during that time; this way the bender can be transferred from another job, eliminating the necessity of purchasing another.

Pre-planning and scheduling can reveal critical situations sufficiently in advance to avoid their consequences and also provide for an orderly servicing and performance of the work in the most efficient manner. But pre-planning and scheduling is not the solution to every problem; it is merely a management tool to provide information that can be helpful in achieving better production support.

From the previous paragraphs, it should be clear that pre-

JOB PROGRESS REPORT

JOB ______________________________ Contract Amount ______________ Sheet No. ______ of ______ Sheets

JOB NO. ______________ Change Orders ______________________________ Date ______________

Dating	MATERIAL		LABOR		DIRECT JOB COST				MISCELLANEOUS		TOTAL COSTS		MAN HOURS	
	Monthly	Total	Monthly	Total	Monthly	Total	Monthly	Total	Monthly	Total	Monthly	Total	Monthly	Total

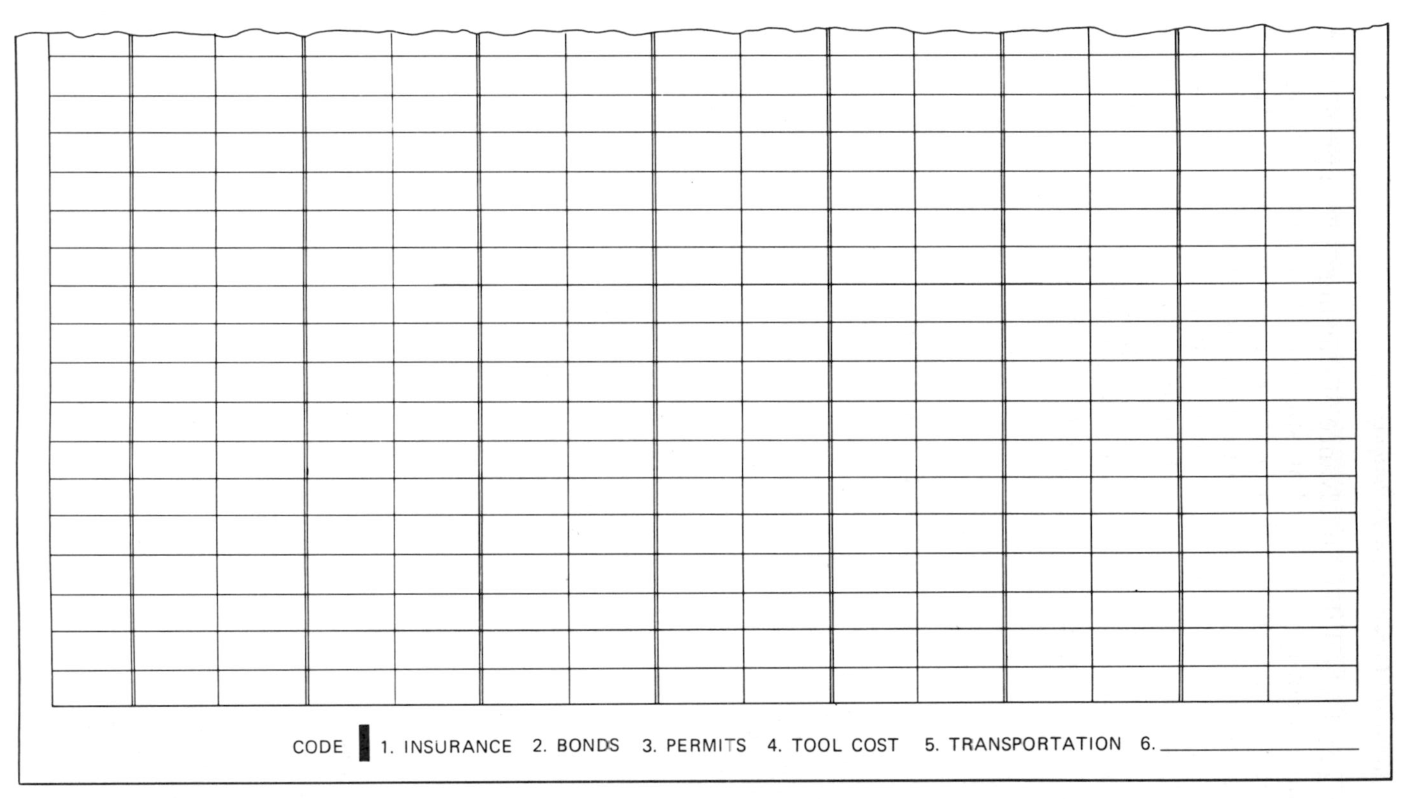

Figure 2.2. Job Progress Report forms are helpful in budgeting performance.

planning and scheduling in advance of starting a given project can save many man-hours of labor spent as lost time waiting for information, materials and tools or else spent through lack of coordination with the general contractor or other trades. In using this system, it is best to follow some orderly plan and sequence. The following check list is therefore designed to help the contractor perform pre-planning and scheduling in the proper sequence. It may also be used to make an approximate determination of the starting and completion dates for the project.

JOB PLANNING CHECKLIST

Contract awarded ____ Date ____
Write up job instructions ____ Date ____
Appoint job supervision personnel ____ Date ____
Obtain plans and specs. ____ Date ____
Review plans and specs. ____ Date ____
Revise plans and specs. ____ Date ____
Prepare additional dwgs. ____ Date ____
Check estimated requirements ____ Date ____
Plan and schedule requirements ____ Date ____
Order shop dwgs. ____ Date ____
Initiate material procurement ____ Date ____
Initiate tool procurement ____ Date ____

Organization Phase	*Date to Activate*
Supervisory personnel to job site ____	____
Provide job shed and office ____	____
Arrange storage facilities ____	____
Initial material delivery ____	____
Initial tool delivery ____	____
Initial workers on job ____	____
Install temporary service ____	____
Install sleeves in footings ____	____
Install grounding in footings ____	____

JOB PLANNING CHECKLIST
(continued)

Performance Phase	*Date to Activate*
Electrical rough-in start ___	___
Electrical rough-in complete ___	___
Overhead and underground system start ___	___
Overhead and underground system complete ___	___
Wire and cable start ___	___
Wire and cable complete ___	___
Panels and other equip. start ___	___
Panels and other equip. complete ___	___
Fixtures and wiring devices start ___	___
Fixtures and wiring devices complete ___	___
Energize and operate system ___	___
Final test and inspection ___	___
Punch list and cleanup ___	___
Job completed and accepted ___	___

Of course this checklist can vary to suit any particular project. On multistory buildings, a separate checklist may be warranted for each floor, especially on the rough-in stage. A more detailed breakdown may be necessary. For example, under "Panels and other equip." you may wish to add subheadings like "motor starter," "dry transformers," "lighting panelboards," "power boards," "main distribution panel."

In any event, when such a plan is incorporated into every project in advance of starting it—before placing expensive labor on the job—less man-hours will normally be consumed than would be expected. In most cases, it will consume no more time of supervisory personnel than is consumed over the period of the job through hit-or-miss operations and more than likely, less time will be involved.

3

Preliminary Job Management

The objectives of this chapter are to point out the various preliminary job requirements prior to moving onto the job site. Such items as planning, scheduling, and ordering necessary materials—at the lowest possible price in line with the required quality and quantity of the item and time of delivery—should be the electrical contractor's first considerations. He should further prepare job progress schedules, appoint job supervision personnel, and begin preparing supplemental working drawings. Planning for temporary electrical service, jobsite shop, and material storage should also be considered at this time.

3.1 Planning, Scheduling, and Ordering Necessary Materials

To provide an efficient basis of material requisitioning, purchasing and follow-up, a complete list of all materials required for a given job should be prepared on a form similar to the one shown in Figure 3.1. Copies should then be routed to the estimating department, purchasing agent, and the job superintendent or foreman, when they are selected.

The project estimator will check the material lists against his take-off and pricing sheets; the purchasing agent will have a complete list of the jobs needs so that firm prices may be obtained as well as definite delivery dates; and the job superintendent or foreman will have a guide as to the types of materials that were planned to be used by the estimator. The job superintendent can also use the list to check the material requirements and any errors or omissions discovered can be corrected to avoid lost time on the job once the project is underway.

If a detailed material take-off was performed by the electrical estimator, it is then usually only necessary to copy the material items onto the Job Material Schedule from the pricing sheets. However, if unit prices were used for the original estimate, it is usually desirable to segregate the various items on the Job Material Schedule. For example, in the case of convenience outlet, the estimator's unit price per outlet may include an outlet box, plaster ring, cable or conduit connector, grounding clip, wiring device with cover, and wire nuts. On the estimate sheet, all of these items may appear as one convenience outlet; on the Job Material Schedule, the items should be broken down to individual items.

Other items to be considered include the smaller wire sizes, where they appeared on the estimate form in total quantities only. On the Job Material Schedule, it would be desirable to list the quantities of wire with different colors of insulation as may be required for identifying branch circuit or control wiring systems.

Once all material items and quantity have been entered in the appropriate place on the form, the date that each item will be required on the job should be provided as well as the price used on the estimate sheets when bidding the job. Stockroom personnel are then notified to determine the quantity of material that is available from inventory which will then give the purchasing agent the quantity of materials that must be obtained elsewhere; that is, from suppliers, manufacturers, and so on.

When it is anticipated that separate related groups of material items will be needed on the job at different times, it is recommended that each such group of items be listed on separate Job Material Schedule forms. This will facilitate the ordering and follow-up of the required items.

3.2 Preparing Job Progress Schedules

The planning and scheduling of electrical projects were briefly discussed in Chapter 2; here a more detailed description is given since this is a very important step in job pre-planning.

In general, budgeting the estimating labor and material requirements of a project begins by obtaining the estimated start and completion dates from the general contractor, and then entering these dates on a progress schedule such as the calendar bar chart in Figure 3.2. Such a chart gives a graphic picture of the

MATERIAL & PRICE SHEET

JOB ______________________ DATE ____________

CONDUITS AND FITTINGS									
KIND	SIZE			SIZE			SIZE		
	QUAN	EACH	TOTAL	QUAN	EACH	TOTAL	QUAN	EACH	TOTAL
Rigid Conduit									
Conduit Ells									
F.E.B.									
L.A.Y.									
LB – L – R									
A – B									
Blank Cover									
Bushings									
Lock Nuts									
Steel Tube									
S.T. Couplings									
S.T. Connectors									
E.M.T. Straps									

OUTLET BOXES–COVERS				SWITCH CABINETS–DEVICES				100 Amp. Cart.			
KIND	QUAN	EACH	TOTAL	KIND	QUAN	EACH	TOTAL	200 Amp. Cart.			
Steel B. Box				30 A Pole Plug							
Steel Bracket				30 A Pole Cart				Bell Wire			
Steel Standard				60 A Pole Plug				Ins. Staples			
Steel Short				60 A Pole Cart				Transformer			
Handy Box				100 A Pole Cart							
Handy Covers				200 A Pole Cart				Rubber Cord			
4-11/16 Sq. $1\frac{1}{2}$								Tape			
4-11/16 Sq. $2\frac{1}{8}$				Range Cab. 4 Cir				BX Staples			
4-11/16 Rings								Cable Straps			
4" Oct. $1\frac{1}{2}$				Breaker Panel							
4" Oct. $2\frac{1}{8}$				Breaker SP				Exp. Bolts			
4" Oct. Rings				Breaker DP				Drive–in Plugs			
4" Square $1\frac{1}{2}$				SP Flush Sw.				Toggle Bolts			
4" Square $2\frac{1}{8}$				SP Surf. Sw.							
4" Sq. Rings				3w Flush Sw.				Connectors			
4" Sq. Covers				4w Flush Sw.							
				Duplex Recpt.							
4" Porc. Pull				We Prf. Recpt.							
								Total			

Conduit Couplings									
Totals									

WIRE									
KIND	NO.			NO.			NO.		
	QUAN	EACH	TOTAL	QUAN	EACH	TOTAL	QUAN	EACH	TOTAL
Rubber Cov.									
Weatherproof									
Non–Met. 2 Wire									
Non–Met. 3 Wire									
Non–Met. w/Grd.									
BX 2 Wire									
BX 3 Wire									
S.E.C. 2 Wire									
S.E.C. 3 Wire									
Lead 2 Wire									
T.W.									
Totals									

Bar Hangers			
Fixture Studs			
Total			
GROUNDING AND HARDWARE			
KIND	QUAN	EACH	TOTAL
Grd. Rods			
Grd. Clamps			
Grd. Straps			
Insulators–Scr			
2 Wire Racks			
3 Wire Racks			
Lag Screws			
Mach. Bolts			
Dead Ends			
Total			

Pilot & Switch			
Sw. Plate–			
Sw. Plate–			
Sw. Plate–			
Recp. Plate–			
Recp. Plate–			
Cabinet–			
Cabinet–			
Total			
MISCELLANEOUS AND FIXTURES			
KIND	QUAN	EACH	TOTAL
Fustats, 1–14			
Fustats, 15–30			
Adaptors			
30 Amp. Cart.			
60 Amp. Cart.			

RECAP	
ITEM	AMOUNT
Conduits & Fittings–	
Conduits & Fittings–	
Conduits & Fittings–	
Wire, No.–	
Wire, No.–	
Wire, No.–	
Boxes & Covers	
Grndg. & Hdwe.	
Sw. Cab's & Dev.	
Misc. & Fixtures	
Total Material	
Labor	
Total	
Overhead	
Profit	
Total	
Sales Tax	
Total	
Sell For–	

Figure 3.1. To provide an efficient basis of material requisitioning, purchasing, and follow-up, a complete list of all materials required for the job should be listed on a convenient form.

DATE WORK TO BE STARTED 11/1/78 : TO BE COMPLETED 80 Calendar

	WEEK / WEEK ENDING		ESTIMATED LABOR		1	2	3	4	5	6	7	8	9	10	11	12	13	14	15	16	17	18
	JOB SECTION OR ITEM DESCRIPTION	CODE	MAN-HOURS	MAN-WEEKS																		
1	Electrical Installation Planning & Scheduling (Initial)																					
2	On Job Make Ready	ORA	600	15		1	1	1	1	1	2	2	2	2	2							
	Temporary Service & Wiring	W7	2,640	66					3	3	2	2	1									
3	Inserts, Sleeves & Hangers	BH	3,410	85					1	1	0	0	0	0	0	1	0	0	1	0	0	1
	Underfloor Duct	DF	15,000	375									1	2	4	6	6	6	18	18	18	18
4	Branch Rough-in)	HR	(16,000)	(400)																		
	✓ Concealed-Slab		4,000	100						1	2	2	2	3	3	3	4	4	3	4	4	3
5	✓ Concealed-Walls & Ceilings		11,600	290	(40 MW-Piece-out In walls: 200 MW Ceiling Rough-in)																	
	✓ Exposed		400	10																		
6	Fixture Rough-in Boxes & Rings	FR	1,800	45												1	1	1				
	Pull Boxes & Troughs	C-4	1,200	30	(Installed Concurrently In Relation To BR Rough-in-Total 43 MW)																	
7	Panel & Cabinet Back Boxes	C-1 & 2	510	13																		
	Feeder Rough-in)	BR	(10,600)	(264)																		
8	✓ Concealed-Slab		620	16																		
	✓ Concealed-Walls & Ceilings		4,360	108	(Installed Concurrently As Continuing Runs-Total 248 MW)																	
9	✓ Exposed		5,620	140																		
	Feeder Busway Risers	BB	480	12	(Receive-Move-in-Install)																	
10	Switch Boards	C-3	940	24																		
	Motor Control Centers	C-5	740	19																		
11	Safety Switches, etc.	C-4	1,480	37																		
	Dry Type Power Transformers	H7	350	9																		
12	Feeder Wire Pull	BW	3,900	97																		
	Branch Wire Pull	AW	5,600	140																		
13	Distribution Panel Interiors	C-2	1,870	46																		
	Branch Panel Interiors	C-1	1,200	30																		
14	(Lighting Fixtures)	F	(23,000)	(575)																		
	Incandescent Fixtures	FI	3,600	90																		
15	Fluorescent Fixtures	FF	13,800	345																		
	Luminous Ceilings & Strips	FL	1,600	40																		
16	Special Chandeliers	FS	1,200	30																		
	Lamps	Flm	1,000	25																		
17	Wiring Devices & Plates	E	1,700	43																		
	Special Systems Equip.	G	1,230	31																		
18	Underfloor Duct Devices	DF	2,100	52																		
	Motors & Controls-Connect	MC	620	16																		
19	Equipment FBO-Connect	EM	1,400	35																		
	Final Inspect-Punch List	ORA	960	24																		
	TOTALS		97,530	2438		1	1	1	5	6	6	6	6	7	9	11	11	11	22	22	22	22

Figure 3.2. Typical calendar bar chart. (Courtesy

overall job from a calendar performance standpoint. The contractor should further enter the start and completion dates of all phases of the project on this chart including the estimated duration of the electrical work. If the general contractor cannot furnish this information, the electrical contractor will have to estimate the conditions himself.

Remember that a given number of man-hours to accomplish a given amount of work has been provided in the estimate and bid. Therefore, if the contractor planned to make his anticipated profit, no more man-hours can be expended on the project.

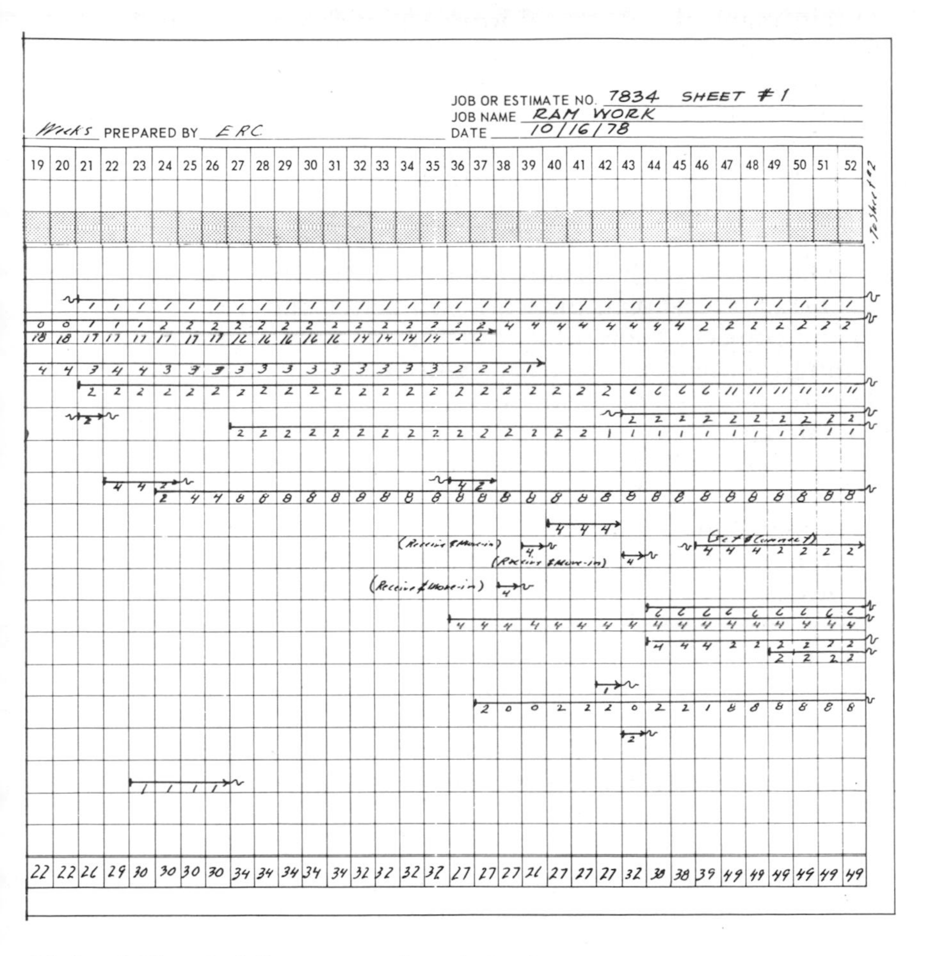

National Electrical Contractors Association.)

On the chart tentatively lay out the calendar durations for each section of the electrical work as related to the various sections of the general construction and those of other subcontractors. In doing so, it will be noted that some sections of the electrical work may not occur for the full duration of a given section of the building construction. For example, during the site layout and excavation, the electrical workers will have little, if any, work on the project. During the concrete construction of the footings, basement walls, and so on, there will still be little electrical work: perhaps the insertion of a few pipe sleeves, and so forth. Then, at a certain point, the electrical work will pick up momentum and often extend over two or more sections of the general building construction.

Once this chart has been roughly laid out, the electrical contractor can divide the total electrical man-hours allowed for each section of the work (as determined from the estimate forms) by the planned electrical work duration in hours to determine the maximum average number of men that can be scheduled for that particular work. With this information at the start of the project, the contractor can plan his workload better and avoid overmanning the project at any time. Such a schedule is also helpful in planning the material needs for the project.

Two or more sections of the electrical work will normally be performed concurrently during certain periods. By analyzing the job situation—utilizing the bar chart—it is often possible to shift certain sections of the electrical work and thereby level out the weekly manload.

This initial planning and scheduling will more than likely have to be adjusted somewhat during the actual construction, but when working under such a plan, labor run-over can be kept to a minimum and often offset completely.

The calendar bar chart just described provides a visual picture of the labor hours estimated; it does not provide for labor actually spent on the project while the job is under construction. Therefore, a comparative bar chart (Figure 3.3) must be provided to show the actual man-hours consumed on each phase of the project, and as compared to the calendar bar chart.

The actual time worked on the project by the various workmen is recorded on some type of daily or weekly time card—one for each worker—for the purpose of preparing the payroll and charging the time to the proper job. The electrical contractor should obtain time cards that allow the segregation of the time worked on the various segregated sections of the job to be recorded separately. Then the contractor can easily transfer the time for each section to the comparative bar chart for analysis.

With respect to the bar chart in Figure 3.3, colored marks may be used more effectively than the solid line bar of one color. One color, for example, can be used to denote the estimated labor arrived from the estimating sheets and the estimated labor required. Another color can be used to denote the actual labor used as the job progresses. A third color may be used next to the actual

Form No. CA-6a: Revised 1964

JOB LABOR PROGRESS CHART
Estimated and Actual Manhours

JOB OR ESTIMATE NO. 6375
JOB NAME Supreme Shopping Center
DATE 4/13/63

DATE WORK TO BE STARTED 4/15/63 : TO BE COMPLETED 3/1/64 PREPARED BY ERC
(Draw Horizontal Bars for Manhours Involved to Appropriate Scale)

Scale: 100 200 300 400 500 600 700 800 900 1000 1100

CODE	DESCRIPTION	MANHOURS	SCALE
AR	Branch Rough-in	1080	Estimated
			Used
AW	Branch Wire & Cable	240	Estimated
			Used
BR	Feeder Rough-in	240	Estimated
			Used
BW	Feeder Wire & Cable	90	Estimated
			Used
X	Hazardous Location Systems	0	Estimated
			Used
C	Panel & Switchboard Equipment	620	Estimated
			Used
D	Special Raceways	40	Estimated
			Used
E	Schedule Material	70	Estimated
			Used
F	Fixtures & Lamps	F 1060 / FS 300	Estimated
			Used
G	Signal System Equipment	0	Estimated
			Used
H	Special Power Equipment	0	Estimated
			Used
I	Incidental Equip. & Appliances	44	Estimated (TM - Equipment FBO Connect.)
			Used
J	Underground Ducts Concrete	0	Estimated
			Used
K	Grounding Systems	0	Estimated
			Used
L	Overhead Systems	0	Estimated
			Used
M	Motors — Controls	0	Estimated
			Used
N		0	Estimated
			Used
O		0	Estimated
			Used
ORA	Move on Site Plan Study	80	Estimated
			Used
WT	Temporary Service	24	Estimated
			Used
TOTAL JOB SECTION		3,888	Estimated
			Used
		SCALE	500 1000 1500 2000 2500 3000 3500 4000

Figure 3.3. Comparative bar chart.

labor used to denote the extent of any labor used in excess of the estimated labor at any given date. The contractor will then know, if the records are accurately kept, almost instantly if any section of the project is in trouble so that steps may be taken to correct it.

Many contractors may think that once a mistake has been made that the damage is done and nothing can be done to rectify it. An example disproves this belief: One electrical contractor was low bidder on a new barracks project at Vint Hill Farms Station (a government installation near Warrington, Virginia). His bid was approximately $23,000 for the electrical work and the closest bid to him was $49,000! The contractor with this seemingly ridiculous bid (as compared to the others) felt certain that he was going to lose a pile of money on this contract, probably enough to cause him to go out of business. Still he took immediate measures to come out of the jam as clearly as possible.

A bar chart was drawn to show the estimated job progress, and the amount of labor that could be expended on each section without running over on his estimated labor. He then had his purchasing agent shop for the best possible material buys; this included looking into possible auctions of electrical contractors or suppliers going out of business with the hopes of obtaining some materials at a very low price.

His best foreman was put in charge of the project, and his best electricians were transferred from other jobs to do the work on this project. Daily records were kept and each night the contractor knew exactly how he stood on the project as far as labor, materials, direct job, and other expenses were concerned.

The job went smoothly and eight months later the project was completed, not at a loss to the electrical contractor, but at a profit of over 12%! Needless to say, the contractor with the bid of $49,000 was high, but the contractor with the $23,000 bid was also too low under normal conditions. However, with careful job planning and supervision, he was able to pull the job out of the hole and make a profit on the installation.

We have been talking mainly about labor expenditures on the bar chart, but a similar chart could be used to show the overall picture of the job condition; that is, labor, materials, direct job expenses, and the like, with a different color bar for each category.

The main point to remember when using a bar chart is to make certain that it be maintained on a current basis and continually referred to in order that a continual picture of the status of progress on the job is kept before those persons responsible for the supervision of the jobs.

Once the electrical contractor or his staff gets into the habit of regularly checking the progress charts of the entire organization, any problem areas will become apparent immediately. Such a chart also provides a psychological affect on some of the workers in that they can keep track of the job progress themselves, and most electrical workers like to make a good showing.

3.3 Appointing Job Supervision Personnel

While most owners and principals of electrical contracting firms are experts in their respective fields, they must also understand the capabilities of their superintendents, foremen, and other employees to be able to select the best personnel to run a given project. Just because one superintendent happens to be capable of supervising, say, residential wiring does not follow that he is well trained and proficient in all phases of electrical work.

The foreman or superintendent who has spent most of his time supervising commercial projects will require time to get adjusted to industrial or large institutional projects. On the other hand, superintendents who have been working mostly on the larger projects usually have even more of a problem adjusting to the smaller ones.

Consideration must also be given to the selection of foremen or "pushers"; a man may be a good electrician for a certain type of electrical installation, but a failure when trying to direct others. Likewise, a foreman may be able to direct three or four men well but unable to control a large crew well enough to obtain maximum efficiency.

Superintendents and general foremen normally are required to have direct contact with the general contractor, architects, engineers, and the owner of the building. Therefore, in selecting personnel for supervision, the tact and diplomacy of the man must also be considered.

Principals of electrical contracting firms not only must have a faculty for selecting new employees but must be able to recognize the abilities of their existing employees as well. They must be familiar with their employees in order to assign them advantageously. This, of course, applies to both new and existing projects.

The electrical contractor who is able to recognize the ability of his employees can often build up an excellent work force out of what may at first appear to be just a group of mediocre electricians and helpers. If at all possible, employees should be assigned to work best suited to them. For example, your most devoted and knowledgeable superintendent may have a fear of height; he could be an excellent superintendent for the majority of your projects, but you wouldn't think of putting him in charge of a high-rise structure.

A great number of electrical workers are good electricians but poor leaders. Others like to take the lead and assume responsibility. Again there are men whose dispositions are such that they cannot be placed in charge of employees. Contractors should take into consideration all of the characteristics of their employees and try to work out the best arrangements for each.

Some contractors may think that the best way to select employees is by hiring and firing until a good work force is built up—a trial-and-error method. Such practice is not only costly, but it is also a way to gain a bad reputation in the industry. Hiring and firing also has a tendency to demoralize the regular employees and may even be responsible for the loss of some customers, should they find out.

Every effort must be made to select the right men during the first hiring, providing that they are available. If not, employees should be selected who are apt, willing to learn, and can readily adapt themselves to the work the contractor normally undertakes or expects to encounter. The experienced electrical contractor can come close to evaluating the characteristics of new employees by merely interviewing them.

An interview usually consists of asking the employee about his educational background, what types of projects he has worked on, for whom, and for how long. Previous employers may or may not be contacted as the experienced contractor or manager usually relies on his own judgment regarding the honesty and ability of the person being interviewed.

Once an electrical contractor has built up a good organization, it must be kept intact. To do this, the managers of the firm must treat their employees—and direct their work—in a manner that will make them desirable employers. Along with the support of projects goes the normal courtesy toward the employees. Each employee is an individual personality and must be regarded accordingly.

Some people have better temperaments than others for supervising employees, but in general, one should be able to control the work rigidly without having to be unpleasant, at least most of the time. The majority of superintendents hired by electrical contracting firms are high-caliber people and like to be made to feel that they are working with their employer rather than for him.

3.4 Preparing Supplemental Working Drawings

The quality of electrical construction documents prepared by architects and their consulting engineers will vary from very excellent to very mediocre, depending upon the training and practical experience of the electrical designer and sometimes upon the money allowed for the engineering services. In some instances, the drawings may be so incomplete that the electrical contractor will have to supplement them before an estimate can be completed or before the project is started.

On even the better electrical drawings, the contractor may wish to supplement the original drawings with more detailed drawings which will help the electricians on the job to understand what is required of them.

Another reason for supplemental drawings is to save the contractor money. Many electrical designers are excellent draftsmen and know electrical engineering, but lack the practical experience necessary to make electrical installations. For this reason, their circuit diagrams are often not drawn to take advantage of the shortest possible runs (to eliminate voltage drop and to save wire and conduit). Therefore, the electrical contractor will often have his draftsman rearrange the circuit routing to save as much material as possible. In this case, the contractor should keep a copy of the changes on file at all times in case the architect or engineer requires "as built" drawings after the project is completed.

Chapter 14 gives additional material on electrical drawings; that is, types of electrical drawings, layout of electrical drawings, electrical graphic symbols and schedules, sectional views, wiring diagrams, site plans, and so on. This chapter will help the electrical contractor interpret electrical drawings prepared by others as well as drawings the contractor desires to produce himself.

3.5 Planning for Temporary Electrical Service

Electrically powered tools are used considerably by almost every craft or trade on the job site, and it is therefore necessary to provide a temporary electric service during the course of construction until the normal power is connected to the building. In most cases, this is the responsibility of the general contractor, although he will usually hire the electrical contractor to do the job for him, either paying the contractor extra for this service or trading services.

When the installation of the temporary wiring is included under the electrical portion of the specifications, it is the responsibility of the electrical contractor to see that the service is installed at the proper time so as not to delay the general contractor on other trades. It is the contractor's responsibility to contact the local power company and arrange for such a service to be installed at

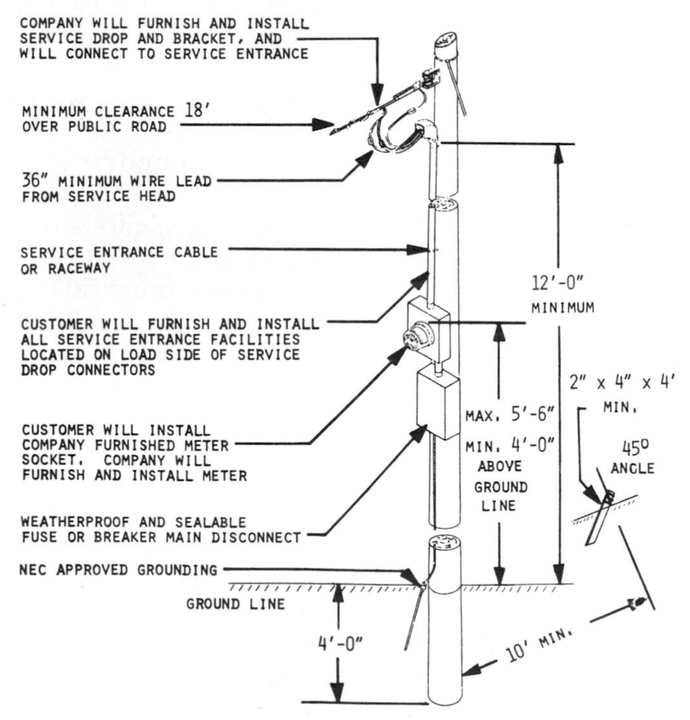

Figure 3.4. A typical single-phase temporary service for residential or small commercial projects.

the job site. This should be done in sufficient time to allow the power company to engineer and construct the needed facilities prior to the in-service or date needed by the general contractor.

Normally the power company will charge a flat fee for providing and connecting temporary service for building construction when providing this type of service can be accomplished by the installation of the service only. The electrical contractor will have to provide an acceptable service support, like a power pole, since the temporary service will not normally be attached to a building that is under construction. The pole must be properly equipped to allow for a convenient and safe installation.

The drawing in Figure 3.4 shows the details of a typical single-phase temporary service for small commercial projects. Note that the power company will furnish and install the service drop bracket, and will connect to the service entrance conductors. This drop must have a clearance of at least 18 feet over public roads and the service entrance conductors must be at least 36 inches in length. Other dimensions are given on the drawing. Note also that a weatherproof and sealable fuse or circuit breaker main disconnect is required as well as grounding according to requirements set forth in the National Electrical Code.

Although not shown, all circuits used on building construction sites must be provided with ground fault interrupters to comply with OSHA standards.

4

Job Management Techniques

This chapter outlines practical and efficient job management procedures as applied to the electrical construction industry. While aimed primarily at those persons with no previous experience in the electrical contracting business, the information should also prove beneficial to those with considerable experience. For such persons, the text will be in the form of refresher or supplementary training.

The main objective, however, is to reduce the cost of operations both to the contractor and the project owners—regardless of the previous experience held by the reader.

4.1 Scope of Job Management

Construction costs must be reduced, and in the face of ever-present wage increases, limitations on man-hour production, and restrictive labor legislation, one way this can be done is to practice good job management. Basically, this means that all tools, materials, and equipment—in the quality/quantity needed—should be on the job site at the time it is required and at the lowest cost. Furthermore, sufficient personnel should be provided to handle and install the materials as specified and in a workmanlike manner.

Material requirements include any item that is installed or used on the job. Such miscellaneous items as wire nuts, tape, wire-pulling lubricants, PVC conduit cement, hack saw blades, small

Figure 4.1. Construction costs must be reduced, and one way this can be accomplished is to practice good job management.

Figure 4.2. The delivery of adequate quantities of the proper items of miscellaneous materials to the job is extremely important to prevent lost time on the job.

Figure 4.3. Such miscellaneous items as wire nuts, tape, and wire-pulling lubricants are just as much a part of the material furnished to the job as conduit and wire, panelboards, and the like.

drills, and wood bits, are just as much a part of the material furnished to the job as conduit and wire, panelboards, lighting fixtures, and the like. All persons concerned with the material servicing of the jobs should be reasonably well acquainted with the many detailed items that are required, to lessen the possibility of overlooking the purchase, requisitioning, and delivery of certain of those components that are necessary to make the installation complete.

The delivery of adequate quantities of the proper items of miscellaneous materials to the job are extremely important to prevent lost time on the job, and money needlessly spent.

Before any materials are ordered, the contractor must first determine:

1. What materials are required.
2. The source of such materials.
3. The price to be paid.
4. When they will be required.
5. How they are to be shipped.

The responsibility for selecting and scheduling materials is best handled by the estimator who prepared the final take-off and pricing. During this process, the estimator has gained a detailed knowledge of the installation and a fairly good idea of the relative sequence in which the various sections of the installation will be made. However, in the case of a completely departmentalized contracting operation where the responsibility for the selection and scheduling of material is assigned to a person other than the estimator who had prepared the estimate, it is necessary for that person to generally review the specifications and drawings and make such job-site investigations as may be necessary to determine the job requirements and check them with the estimate pricing sheets before selecting the materials. It is necessary to do this in order to keep the job costs in line with the estimate cost allowances. In the case of this type of operating a high degree of cooperation between the person doing the selection and scheduling work and the estimator is necessary.

The sections of the job and the items of materials normally listed together are as follows (see also Figure 4.4):

Branch circuit wiring consists of such items of material as all outlet boxes of all types and their covers, all ½-inch, ¾-inch, and 1-inch conduit and fittings, and all types of building wires up to and including No. 8. This section will also include all sheathed cable up to and including No. 8, and will normally include all wiring for lighting, receptacles, small power, and signal systems.

Service and feeder wiring, including heavy power branch circuits, covers all 1¼-inch and larger conduits and fittings; fastenings and hangers and all types of cable from and including No. 6 and larger. This section will also include sheathed power and feeder cable of No. 6 and larger. Grounding material would also be included in this section. (See Figure 4.5.)

Panelboard and switchboard equipment includes all panelboard, switchboards, metering equipment, externally operated switches and circuit breakers, auxiliary gutters, pull boxes, telephone cabinets, motor control centers, individual motor starters, and so on, together with the necessary accessory items, hangers and backing; that is, all of the equipment normally supplied by a switchboard manufacturing company, often under a lump-sum quotation. (See Figure 4.6.)

Special raceway systems include the materials and accessory items for all types of raceway systems other than rigid conduit, electric metallic tubing and flexible conduit. They are items that

Figure 4.4. Branch circuit consists of such items of material as outlet boxes, extension rings, all ½-inch through 1-inch conduit and fittings, and all types of building wires up to and including No. 8 AWG.

Figure 4.5. Service and feeder wiring, including heavy power branch circuits, covers all 1¼-inch and larger conduits and fittings.

Figure 4.6. Panelboard and switchboard equipment includes all panelboards, . . .

. . . metering equipment, . . .

. . . externally operated switches, . . .

. . . pullboxes, and similar items.

require special outlets and fittings, and so on, such as metal molding and surface duct, as well as underfloor duct and cellular floors. (See Figure 4.7.)

Schedule material includes wall switches, receptacles, flush plates, bells, buzzers, bell pushes, and so forth. Because of the sequence of shipment to, and installation on the job, fuses usually should be included in this category.

Lighting fixtures and lamps include any lamp-holding receptacles or lighting fixtures including the lamps when furnished, as well as accessory operating equipment, supports, and so on.

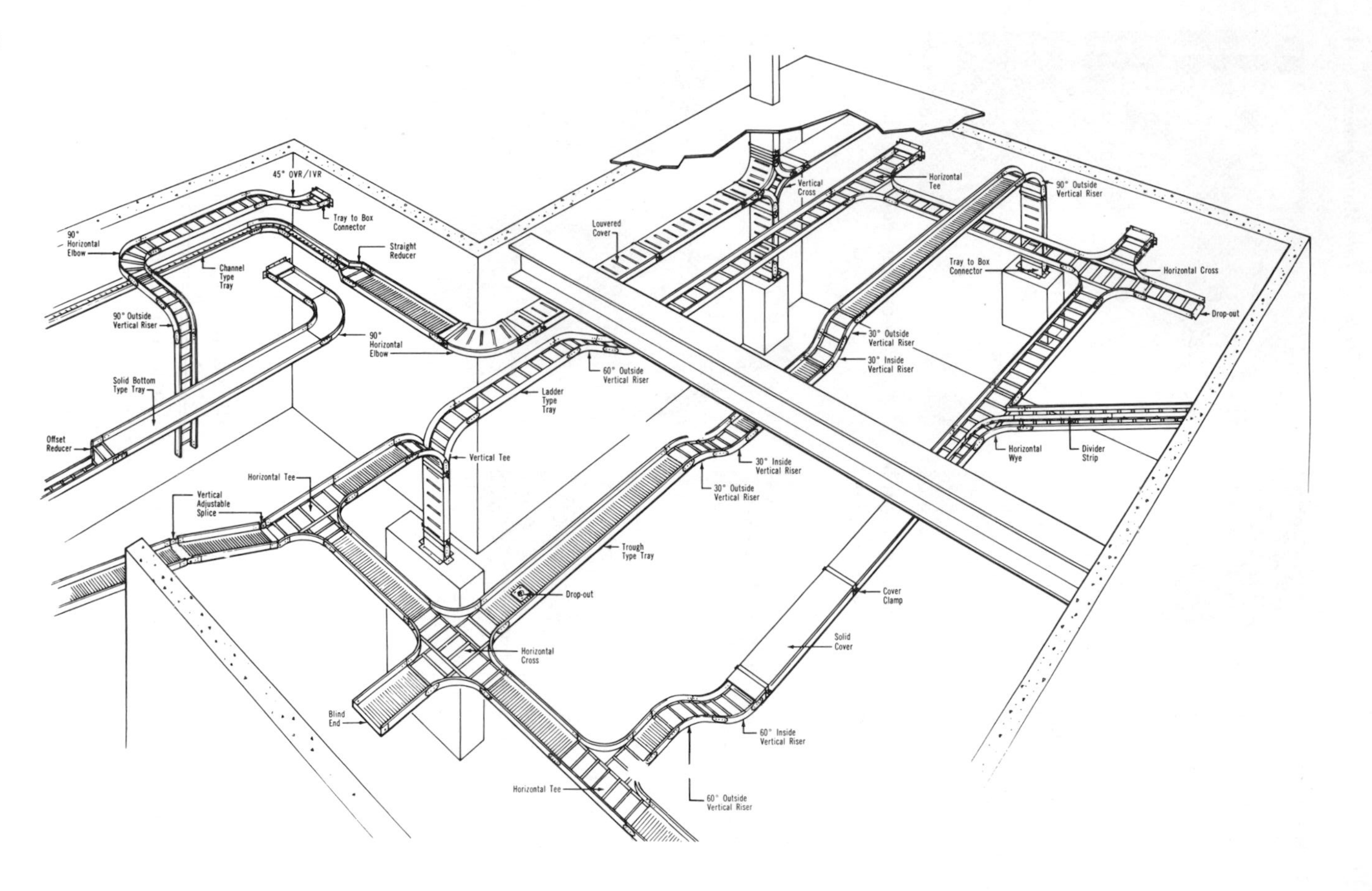
45° OVR/IVR
Tray to Box Connector
90° Horizontal Elbow
Channel Type Tray
Straight Reducer
90° Outside Vertical Riser
90° Horizontal Elbow
Solid Bottom Type Tray
Offset Reducer
Louvered Cover
Vertical Cross
Horizontal Tee
90° Outside Vertical Riser
Tray to Box Connector
Horizontal Cross
Drop-out
30° Outside Vertical Riser
30° Inside Vertical Riser
60° Outside Vertical Riser
Ladder Type Tray
Vertical Tee
Horizontal Tee
Vertical Adjustable Splice
Horizontal Wye
Divider Strip
30° Inside Vertical Riser
30° Outside Vertical Riser
Trough Type Tray
Drop-out
Cover Clamp
Solid Cover
Horizontal Cross
Blind End
60° Inside Vertical Riser
Horizontal Tee
60° Outside Vertical Riser

Communication and signal equipment include all items of equipment comprising a completely integrated signal system. Some estimators may desire to include signal system wiring under this section rather than under branch circuit wiring.

Special power equipment includes transformers, transformer vault equipment, switchgear, unit substations, capacitors, motors, motor generator sets and their control equipment, and other large power equipment. Bus duct or trolley duct may be included under this section. This section should also include the labor for connecting all motors, controllers, and other equipment furnished by others, but installed or connected under the electrical contract, indicating miscellaneous materials required.

Special incidental equipment and appliances include electric air heaters, hot water heaters, ranges, fans, clocks not a part of a signal system, and all other appliances included in the electrical contract. This section should also include the labor for the connecting of all such appliances furnished by others and connected under the electrical contract.

Underground ducts, trenching, and concrete include all ducts and their fittings, as well as all trenching and concrete envelope for any purpose. It also includes manholes, manhole hardware, concrete pull boxes; also street, highway and airport lighting concrete bases, and transformer and substation concrete pads.

When the proper materials are not available on the job at the time they are required, unnecessary extra labor costs may be incurred because of lost time in waiting for the materials or in shifting men to other jobs. This condition may also delay the final completion of the job. Any extra charges (penalties) are charged against the electrical contractor.

On the other hand, when materials are received far in advance of the time that they will be needed, working capital is needlessly tied up, and added costs of interest on bank loans to pay for the materials may be incurred. Extra handling and storage costs may also be incurred as well as costs due to the possibility of breakage or damage.

The normal sequence of installing electrical systems for wiring

---

Figure 4.7. Special raceway systems include the materials and accessory items for all types of raceway systems other than rigid and flexible conduit. One system falling under this category would be a cable tray system shown here: Cable Strut® cable tray with continuous rigid cable support.

Figure 4.8. Special power equipment includes transformer vaults.

within building structures and hence the sequence of material requirements is as follows:

1. Temporary wiring.
2. Branch circuit conduit or cable systems for lighting and power circuits as well as signal systems including all related components.
3. Larger sizes of conduit or cables.
4. Switchboards and switchgear.
5. Wires and cables.
6. Lighting fixtures and lamps.
7. Communication and signal systems.
8. Special power equipment.
9. Special incidental equipment and appliances.

4.2 Requisitioning

There are normally two basic sources of materials, either from the contractor's inventory stock or from some outside source such as a wholesaler or manufacturer. In planning deliveries from inventory stocks, the requirements for fill-in deliveries to large jobs and the complete requirements for current small jobs must be taken into consideration in order that the inventory stocks not be depleted to the point that such deliveries could not be made.

In order to provide a basis of cost accounting for all materials used on firm contract jobs as well as a basis for invoicing and cost accounting for time and material and cost plus jobs, a record of all materials taken from inventory stock must be made, as well as a record of all unused materials returned from the job to inventory stock. Such a record is most efficiently and accurately maintained by using some type of material requisition form such as the one in Figure 4.9. The main purpose of this requisition form is to definitely establish the fact that the material has been delivered to and received on the job. Therefore, the delivery of all material to a job should be acknowledged by someone having responsibility like the job superintendent, foreman, or electrician.

Further purposes of material requisitions are:

1. To provide a basis of charging the job with the cost of the items delivered and preparation of any necessary invoices.
2. To provide a record on the job of materials delivered from the shop or inventory.
3. To provide a record of the extent of deliveries made from inventory.
4. To provide necessary information to warehouse personnel to make the required deliveries.

A form such as the one in Figure 4.1 should be used every time materials and tools are taken from stock. At least three copies should be made (additional copies may be desirable or necessary for the firm's requisition procedure) and these should be routed to the stock room or warehouse so that the materials may be assembled for delivery.

The quantity of each item should be entered in the left-hand column of the form; the second column from the left is provided for a job or code number, while the next column is provided for a description of the items; that is, wire, outlet boxes, and so on. The actual quantity shipped from inventory as well as any quantity back-ordered is entered on all three copies of the form also.

Once the entire shipment is assembled, the original copy of requisition form is held in the stockroom and the remaining two copies should accompany the materials to the job site.

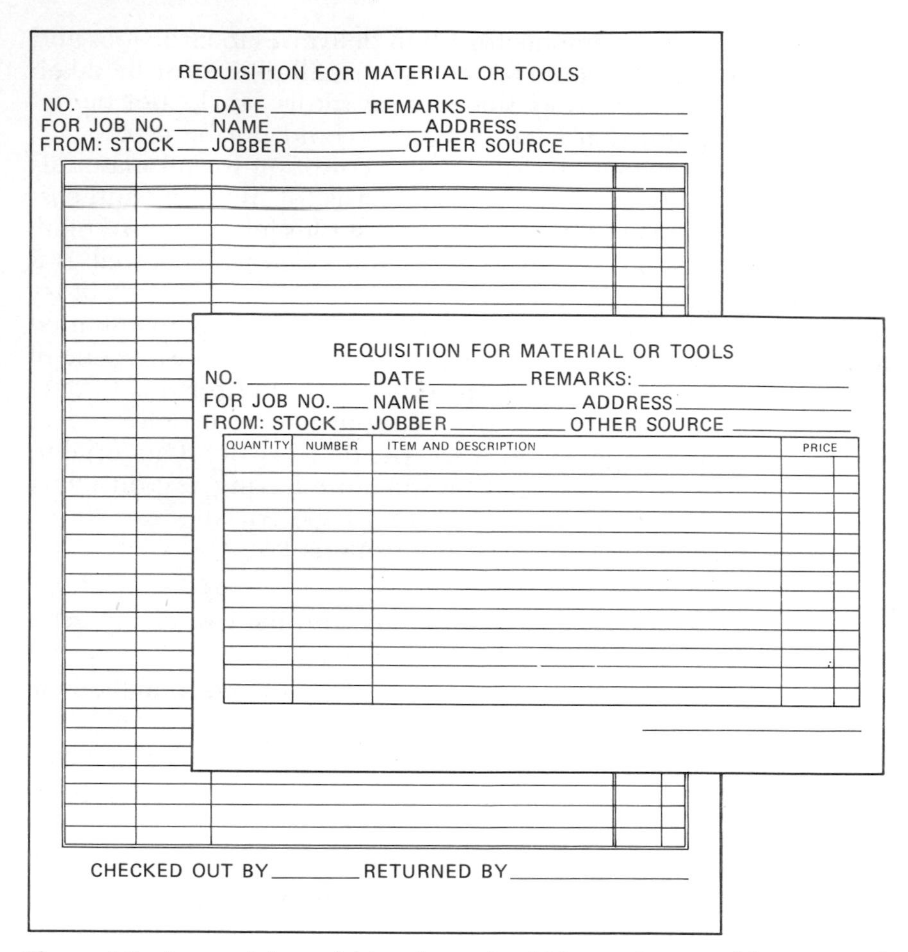

REQUISITION FOR MATERIAL OR TOOLS

NO. ______ DATE ______ REMARKS ______
FOR JOB NO. ______ NAME ______ ADDRESS ______
FROM: STOCK ______ JOBBER ______ OTHER SOURCE ______

CHECKED OUT BY ______ RETURNED BY ______

REQUISITION FOR MATERIAL OR TOOLS

NO. ______ DATE ______ REMARKS: ______
FOR JOB NO. ______ NAME ______ ADDRESS ______
FROM: STOCK ______ JOBBER ______ OTHER SOURCE ______

QUANTITY	NUMBER	ITEM AND DESCRIPTION	PRICE

Figure 4.9. A material requisition form should be used to provide a basis of cost accounting for all materials used on contract jobs.

When the delivery is made at the job site, the copies should be signed by a responsible person on the job. One copy is retained on the job and the other copy goes back to the stockroom or warehouse where it is held until all back orders are filled. Then the original and another copy (the one signed by the job personnel) is routed to the accounting department where it is held until costing of the requisition and any necessary invoicing and stock control recording is completed. After the necessary accounting procedures are completed, the original copy may be placed in a master requisition file while the other copy always goes into the job file for general reference.

Excess materials are often delivered to the job site. If a particular phase of the job is completed and the job superintendent is reasonably certain that the remaining materials will not be needed on the project, they should be returned to inventory as soon as possible so that the excess materials may be used on another project. This keeps the value of the inventory stock to a minimum and prevents the unnecessary purchase of new items.

4.3 Purchasing

Wise purchasing can contribute greatly to the profitmaking potential of the electrical contracting firm. In fact, the purchasing of materials and the proper follow-up on their delivery is one of the most important phases of job management.

In general, purchasing of materials is either for replenishing inventory stock or for direct job use. In most cases, materials are delivered directly to the job site unless storage facilities are limited and the materials will not be needed immediately. In this case, the materials are normally shipped to the stockroom or warehouse where they will be held for future delivery to the job site when needed. However, the contractor should strive to have all materials ordered for a particular job delivered directly to the job site. This eliminates much handling and thus saves money in most cases.

During the course of estimating the project, material prices are usually obtained from several suppliers and manufacturers, with the lowest bidder getting the purchase order for the materials. Some contractors, however, have used the lowest lump sum material prices as a basis for their bids, but then shopped around for better prices once the contract was awarded. Their attitude is that, just as many general contractors shop around for their subcontractors, the electrical contractor can shop around for better material prices. The fact is that an electrical contractor must be a sharp, close buyer, but he also must be fair to his suppliers. If the salesman spent days and nights sharpening his pencil trying to get the overall material cost for your project down to the bare minimum, and using his prices, you get the job, it's only fair that you give him the order for the material.

Some contractors even go so far as to take the lowest material price received from one supplier, and ask other suppliers—after the job is let—to see if they can beat any of the prices. Those who practice this procedure will soon find the suppliers helping other electrical contractors to obtain work instead of his company.

Once the supplier has been selected to furnish a particular job with materials, a purchase order should be submitted for all orders. If the order is given over the phone, a purchase order number should be assigned to the order and a formal purchase order prepared and mailed to the supplier. However, this written purchase order (following a verbal order) should clearly state that it is a confirming order, to avoid duplicate shipments.

If the superintendant orders materials directly from the job site, he should first call the main office and obtain a purchase order number. On the same order, if material is picked up at the supplier's place of business, a purchase order number must be assigned to the order as well as detailed information as to what project the material should be charged against.

Figure 4.10 shows a sample purchase order form. This typifies a good purchase order that relates to all the information that should be shown on a purchase order. For example, it has provisions for designating the name and address of the supplier, the date of the order, the purchase order number, and a simple set of coded instructions to make the order clear. This form further has spaces to designate the exact quantity and description of each item ordered, the price quoted, and the price the contractor expects to pay as well as the terms of payment.

Another easily read purchase order is shown in Figure 4.11. This particular form is available in snap-out style and comes in duplicate or triplicate. Most contractors have the firm name imprinted on purchase order forms and also the purchase order number.

Regardless of the type of purchase order the contractor may use, the purposes are the same; that is, to furnish the necessary information to the supplier and to provide a basis for checking deliveries to the job site and to the stockroom. The purchase order will further provide a means of checking invoices, a basis of charging the invoice amount to the accounting and job cost records, to indicate the extent of purchases for specific jobs, and to provide information for inventory control.

Three copies are normally used for the conventional purchase order. The original is sent to the supplier, one copy is placed in

PURCHASE ORDER

Please Show this Number on Invoice and Shipment

No. ____________

SHIP TO: ________________________________

____

____

TO: ________________________________

____

____

Date ________________________

INSTRUCTION CODE

BO–Back Order if not in stock and ship as soon as available.

C – Cancel and advise if not able to ship immediately.

D – Direct Shipment – order from factory if not in stock.

P – Pick up from other suppliers if possible.

Q – Quote before you make shipment.

SHIP WHEN ____________ VIA ________________

On Future Shipment Orders ☐ Notify First if Prices Advance ☐ Ship at Best Price Available

Quan	Unit	Stock No.	Description	List	Net or discount	Total		Code

TERMS ________________________ SIGNED ________________________

________________________ BY ________________________

Figure 4.10. Sample purchase order.

PURCHASE ORDER

This Order Number Must Appear on All Invoices, Packages, etc.

FROM:

DATE: __________

SHIP TO: __________

TO: __________

SHIP VIA:	DATE REQUIRED:	TERMS:

Item	Quantity	Catalog No.	Description	Unit Price	Amount
1					
2					
3					
4					
5					
6					
7					
8					
9					
10					
11					
12					
13					
14					
15					
16					
17					
18					
19					
20					
21					

Please send _____ copies of your invoice.
Please Acknowledge

By __________
Authorized Signature

Figure 4.11. Another type of purchase order.

numerical order in a master purchase order file, and the other copy is placed in the appropriate individual job file.

When the material reaches the job site or the stockroom, the shipment should be checked to determine the exact quantity received, if the correct items were shipped, and if there is any damage or imperfections.

4.4 Material Control, Handling, and Storage

The extent of an electrical contractor's inventory must be in accordance with the firm's size and the number and type of projects normally encountered. The inventory stock cannot be too large, since it can impair the firm's working capital. Neither can it be too small, because this will hamper the servicing of smaller jobs as well as the larger projects that need urgent fill-in orders to eliminate or reduce lost time when the need arises.

In determining the extent of the contractor's inventory, the following factors should be considered: (1) the general availability of materials in the contractors area, (2) normal delivery time required, and (3) the type of work most performed by the contractor. A contractor doing small projects, for example, would require a relatively large inventory to service the many small jobs, while a contractor doing only a few large projects each year, with most materials being shipped directly to the job site, would need only a small inventory stock which probably could be had by material left over from the jobs.

The storage of materials is another important phase of job management. The material should be arranged in a stockroom or warehouse so that the materials can be handled with a minimum of effort—the least amount of shifting around and extra handling. All storage bins and shelves should be readily accessible with no equipment blocking access to them. Of course, adequate floor space should be provided to handle the amount of material required for the contractor's normal operation and, if possible, all facilities should be located on one floor.

Handling materials on the job site is equally important to the storage and handling of material in the stockroom or warehouse. On most of the larger jobs, a general material storage area is set up either in a trailer or a temporary building constructed by the contractor for the storage of material. Once the job is underway, it is customary to stockpile materials at various locations on the job near the area where the material will be used to decrease the time required to obtain needed materials.

The general storage area should be arranged so that the materials are stored in an orderly manner; that is, all materials are readily accessible and quantities on hand can easily be checked when necessary. One or more trailers that can be moved from job to job seem to be the most popular method of storing materials on construction sites due to the high cost of constructing temporary buildings. However, on some projects where a certain amount of the building is under roof before too much electrical work is required, many contractors use a portion of the building as the material storage area.

For example, during the construction of many high-rise buildings, one or two parking levels are normally completed before any great number of electricians are required on the project. During these preliminary stages, conduit and fittings may be stored outside while tools and items that may "walk off" the job are stored in gang boxes secured with padlocks. Then when the parking levels are completed, the electrical contractor could partition one portion of the area to store material throughout the remainder of the project.

Many new buildings have basement rooms which may be locked during the early stages of construction. These rooms also make good material storage areas during the duration of the project. On a job of any size, temporary shelves and bins should be provided, rather than piling the materials on the floor and on top of each other. In any event, the material storage area should be kept locked when not under surveillance of the electricians or supervisors to avoid pilferage of materials.

When stockpiling materials near the point where they will be used—to reduce the amount of time taken by workers in going to and from the central storage location—large watertight gang boxes will suffice for the storage of the smaller items. For the larger items, a room or closet will probably hold them. During construction, hardware other than hinges will probably be omitted until the final stages of the construction. However, a few dollars spent on a knob and lock set for one or two doors which will offer security for your material room is much better than laying out much more cash for a job trailer or having an area partitioned off.

Conduit and similar materials should be stacked or piled away from other building construction materials and should be prevented from coming into contact with the ground. On very large projects, conduit and fittings of the required amount and sizes are usually stored on each floor of the building to eliminate the need of carrying a supply to each floor daily. A good electrical superintendent will arrange to have the conduit lifted by the construction crane during the construction of each level.

Other materials besides conduit may be the responsibility of the foremen in that each foreman will be required to maintain a sufficient amount of the required materials near his work, and in advance of their need. On the larger projects covering large areas, it is advisable to provide small movable materials cribs in which the smaller items of material may be temporarily stored.

The extra time taken to arrange for all of these conveniences will more than offset the cost by cutting installation time considerably.

4.5 Selecting and Scheduling Tools

The proper selection and scheduling of tools is highly important to the electrical contractor's operation since tools and equipment are the means which are used to combine the materials and operating equipment, along with the labor, into a finished electrical installation. The furnishing of such tools requires a relatively large investment, not only to make the initial purchase, but also to maintain and repair them.

It has been proven that the use of proper tools on electrical installations will lessen the amount of labor consumed for any given installation. For this reason, it is very important that the contractor provide tools of the type and in the quantity and condition that will allow the installation to be made with the use of the minimum number of man-hours of labor.

Over the past couple of decades, the use of power operated tools of one form or another has all but replaced the common hand tools of yesteryear. These power tools have enabled electrical installations to be completed in a much less period of time; their use also reduces worker fatigue and, in some cases, improves the overall quality of the work.

Union agreements vary from local to local, but in most cases, the electrical worker will be required to furnish the following tools:

1 – Tool box	1 – Pr. 9″ side cutters
2 – Pr. channel locks (pliers)	2 – Screwdrivers
1 – 6′ Rule	1 – Voltage tester
1 – Center punch	1 – Level (about 8″)
1 – Flashlight	1 – Adjustable end wrench
1 – Claw hammer	1 – Hacksaw
1 – Knife	1 – Tap wrench
1 – Pr. diagonal pliers	1 – Pr. long-nose pliers
1 – Fuse puller	1 – Lock

An experienced electrical worker usually will have many more items in his toolbox than those required by the union agreement. They will be hand tools which the electrician personally likes to use and which are an aid to performing his work more efficiently and in a more workmanlike manner and with less fatigue. Good tools cannot make a good electrician, but more often than not, an electrical worker can be judged by the type and condition of hand tools which he carries in his tool box.

All other tools not listed in the labor agreement are furnished by the electrical contractor. In general, all tools furnished by the contractor will fall into one or more of the following categories:

1. Shop tools and equipment.
2. Expendable tools (drill bits, etc.).
3. Conduit and raceway tools.
4. Wire and cable installation tools.
5. Cable splicing tools.
6. Trucks and other automotive equipment.
7. Testing and measuring equipment.
8. General construction tools and equipment.
9. Portable power tools.
10. Tools for special systems.

Few electrical contractors purchase outright all the tools and equipment necessary at one time. Rather, tools are normally purchased or rented on an "as needed" basis. An established contractor, however, generally has sufficient tools and equipment to perform the types and sizes of work which he normally undertakes to perform. He has found out long ago that tools are a good investment for promoting efficiency and saving time and money on any electrical contract.

The quantities of tools to be purchased to properly service the contractor's projects will depend upon the relative life of the tools in question, the number and size of the jobs, length of time allowed to perform the work, and the relative location of the jobs.

To illustrate, tool items having a relatively short life such as drill bits would be purchased in sufficient quantity that worn bits could be replaced promptly as needed. Some shops have the worn bits returned to the shop for sharpening; others merely discard them as the labor to resharpen bits is more costly than a new bit. The larger the number of jobs being performed, the larger the number of basic tool items required for servicing each job. Likewise, a project that must be performed quickly requires more men on the job, which means more tools. When several jobs are being performed in different locations causing a longer delivery and pick-up time, a greater quantity of tool items would have to be maintained and purchased.

In the case of certain tools that are used infrequently, the contractor may be better off renting the equipment rather than purchasing it. In fact, in some areas, electrical contractors have found it advantageous to establish a tool and equipment pool, particularly for the less frequently used items of equipment which would involve a considerable investment were each contractor to maintain a complete stock to serve infrequent or peak requirements.

The advantages of such an equipment pool are to make such equipment readily available to contractors not having it in their own stock, or to reduce the necessary capital investments of all contractors and to recompense the contractors having such equipment in their stocks for their capital investment and maintenance costs.

The problems with such an equipment pool are misuse of the tools or equipment by the renter, accidental breakage, pick-up and delivery, and sudden need by the contractor supplying the equipment when it would be a disadvantage to the contractor using it to release it.

Electrical contractors should keep the ratio of capital invested in tools and installation equipment to the volume of business at a minimum consistent with properly servicing the work. Over-investment in such items can seriously jeopardize their capital structure as well as cash position in paying for materials and meeting payroll requirements. Personal property taxes are often assessed on such items as the contractor's tools (in the case of a single ownership type business). This can introduce an overhead expense inconsistent with the volume of work being performed if the tool investment to volume of work ratio is too high.

A contractor performing less than $100,000 worth of electrical work each year should not have more than 4% or $4000 invested in tools. As the volume of business rises, the percentage of tool cost to volume of business reduces.

5

Basic Installation Practices

This chapter covers the more specific installation practices—as applied to the electrical contractor—of building construction. Factors causing excessive labor are the prime consideration since this is the main reason for exceeding the contract amount on most projects.

A further objective is to outline some of the improved installation practices which contribute to reducing the labor cost and, in many cases, the cost of materials on electrical installations.

5.1 Factors Causing Excessive Labor

Prior to performing any installations on a project of any size, proper working drawings or layout instructions must be provided. The lack of such instructions is one of the greatest causes of excessive labor on electrical construction work.

Improperly designed drawings or those with crowded or confusing lines, symbols, and notes are just as bad as no drawings at all, maybe even worse. When the engineered drawings are not complete, or the layout is crowded or confusing, the electrical contractor should have drawings of his own produced by his drafting or engineering department. If the design is incorrect, corrections should be made in red pencil on the original drawings, and then given to the drafting department for redrawing.

Crowded or confusing drawings may be improved by having the drawings made to a larger scale using improved electrical

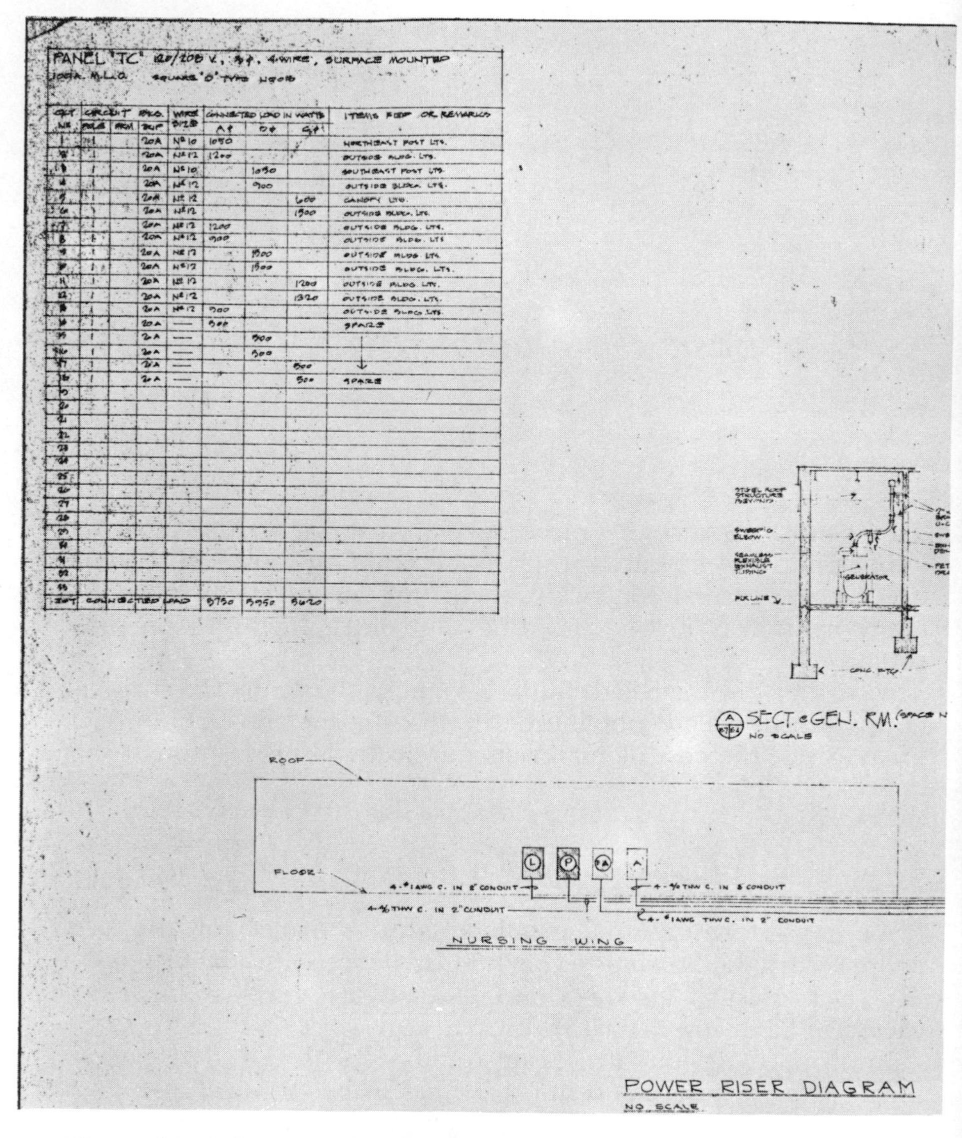

Figure 5.1. Prior to performing any installations on a project of any size, proper working drawings or layout instructions must be provided.

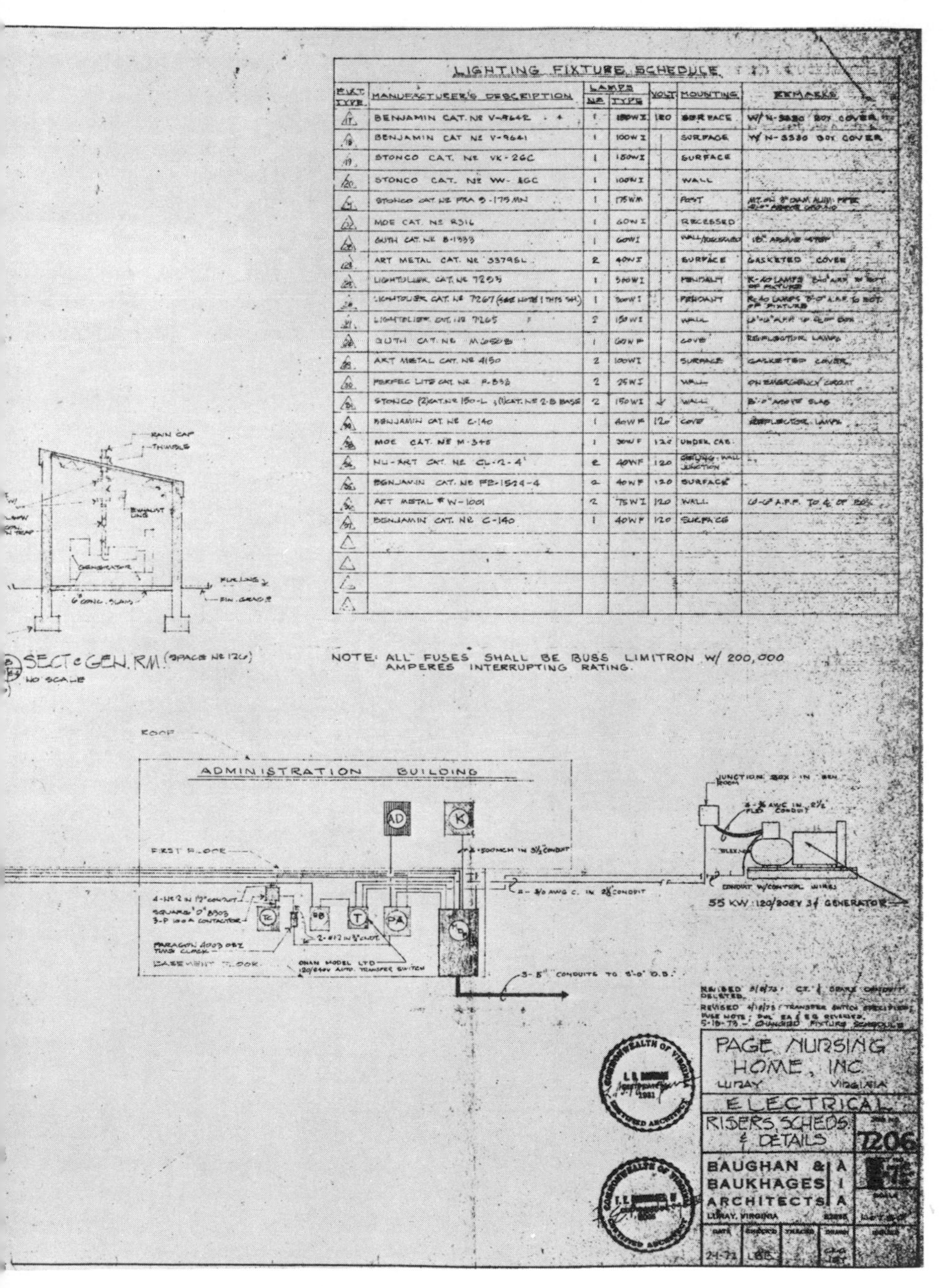

Figure 5.1 (Continued)

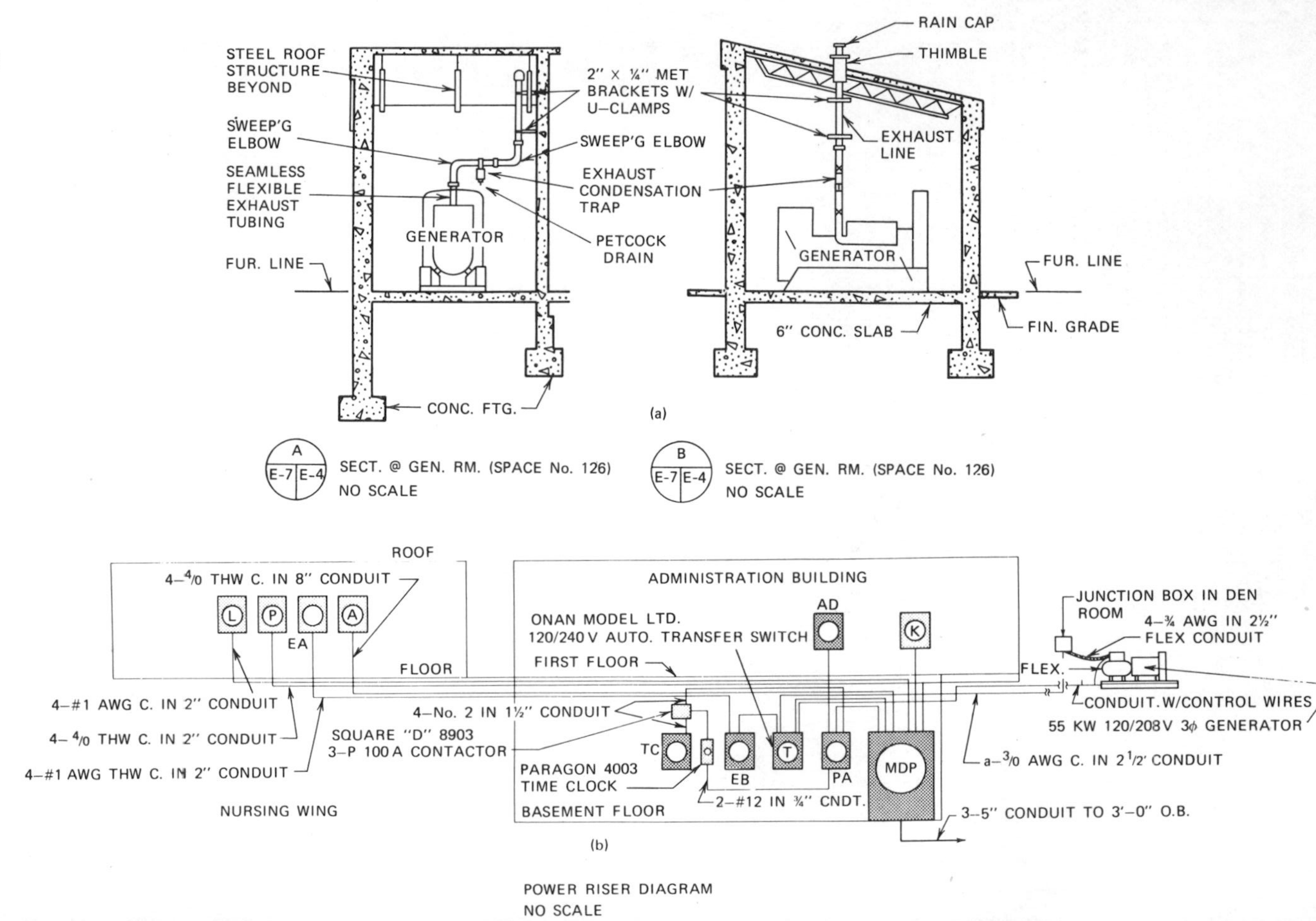

POWER RISER DIAGRAM
NO SCALE

PANEL "TC" 120/208V., 3ϕ, 4-WIRE SURFACE MOUNTED
100A A.L.O SQUARE "D" TYPE HQOB

CKT No.	CIRCUIT BKS.			WIRE SIZE	CONNECTED LOAD IN WATTS			ITEMS FED OR REMARKS
	POLE	FRM	TRIP		Aϕ	Bϕ	Cϕ	
1	1		20A	No. 10	1050			NORTHEAST POST LTS.
2	1		20A	No. 12	1200			OUTSIDE BLDG. LTS.
3	1		20A	No. 10		1050		SOUTHEAST POST LTS.
4	1		20A	No. 12		900		OUTSIDE BLDG. LTS.
5	1		20A	No. 12			600	CANOPY LTS.
6	1		20A	No. 12			1500	OUTSIDE BLDG. LTS.
7	1		20A	No. 12	1200			OUTSIDE BLDG. LTS.
8	1		20A	No. 12	900			OUTSIDE BLDG. LTS.
9	1		20A	No. 12		1500		OUTSIDE BLDG. LTS.
10	1		20A	No. 12		1500		OUTSIDE BLDG. LTS.
11	1		20A	No. 12			1200	OUTSIDE BLDG. LTS.
12	1		20A	No. 12			1320	OUTSIDE BLDG. LTS.
13	1		20A	No. 12	900			OUTSIDE BLDG. LTS.
14	1		20A	——	900			SPARE
15	1		20A	——		500		↓
16	1		20A	——		500		↓
17	1		20A	——			500	↓
18	1		20A	——			500	SPARE
TOT.	CONNECTED LOAD				5750	5950	5620	

(c)

LIGHTING FIXTURE SCHEDULE

FIXT. TYPE.	MANUFACTURER.S DESCRIPTION	LAMPS		VOLT	MOUNTING	REMARKS
		No.	TYPE			
17	BENJAMIN CAT. No. V-9742	1	150W I	120	SURFACE	W/N-3380 BOX COVER
18	BENJAMIN CAT. No. v-9641	1	100W I	│	SURFACE	W/N-3380 BOX COVER
19	STONCO CAT. No. VK-26C	1	150W I	│	SURFACE	
20	STONCO CAT. No. VW-16C	1	100W I	│	WALL	
21	STONCO CAT. No. PRA 9-175 MN	1	175 W/M	│	POST	MT. ON 3" DIAM ALUM. PIPE 4'-0" ABOVE GROUND
22	MOE CAT. No. R316	1	60W I	│	RECESSED	
23	GUTH CAT. No. B-1333	1	60W I	│	WALL/ RECESSED	18" ABOVE STEP
24	ART METAL CAT. No. 3379SL	2	40W I	│	SURFACE	GASKETED COVER
25	LIGHTOLIER CAT. No. 7293	1	300W I	│	PENDANT	R-40 LAMPS 8'-0" A.R.P. TO BOT. OF FIXTURE
26	LIGHTOLIER CAT. No. 7267 (SEE NOTE: THIS SH.)	1	300W I	│	PENDANT	R-40 LAMPS 8'-0" A.F.F. TO BOT. OF FIXTURE
27	LIGHTOLIER CAT. No. 7265	2	150W I	│	WALL	6'-6" A.F.F. TO ℄ OF BOX
28	GUTH CAT. No. M6508	1	60W F	│	COVE	REFLECTOR LAMPS
29	ART METAL CAT. No. 4150	2	100W I	│	SURFACE	GASKETED COVER
30	PERFEC LITE CAT. No. P-833	2	25W I	│	WALL	ON EMERGENCY CIRCUIT
31	STONCO (2) CAT. No. 150-L (1); CAT. No. 2-B BASE	2	150W I	↓	WALL	8'-0" ABOVE SLAB
32	BENJAMIN CAT. No. C-140	1	40W F	120	COVE	REFLECTOR LAMPS
33	MOE CAT. No. M-348	1	30W F	120	UNDER CAB.	
34	NU-ART CAT. No. CL-2-4'	2	40W F	120	CEILING-WALL JUNCTION	
35	BENJAMIN CAT. No. FE-1524-4	2	40W F	120	SURFACE	
36	ART METAL No. W-1001	2	75W I	120	WALL	6'-6" A.F.F. TO ℄ OF BOX
37	BENJAMIN CAT. No. C-140	1	40W I	120	SURFACE	

NOTE: ALL FUSES SHALL BE BUSS LIMITRON W/200,000 AMPERES INTERRUPTING RATING.

(d)

Figure 5.1 (Continued)

symbols. It is also often advantageous to have the different sections of the electrical system on separate sheets of drawings. For example, the power wiring should appear on one drawing and the lighting plan on another. If the drawings are still crowded, further segregation can be made by separating the signal systems, feeders, and so on.

Many times electrical drawings prepared by draftsmen or designers do not take full advantage of the most efficient combination of circuits and homeruns. If the electricians on the job follow these instructions, more labor and material cost will accrue than if the circuits were rerouted in a more efficient fashion. Superintendents and foremen normally catch these deficiencies during the rough-in period of the project. However, additional supplemental drawings prepared by the contractor's design/drafting department will insure that such conditions are not overlooked; layout time on the job will also be saved.

In some instances, when extremely accurate outlet locations are required, much on-the-job time can be saved when supplemental drawings of a larger scale and with exact dimensions are provided. Many contractors may think that the extra cost of providing the additional drawing will be too costly. While it is true that drawings are expensive, the savings from reduced labor cost will usually affect such costs.

Electrical designers rarely indicate the exact types of outlet boxes to be used in a given area, with the possible exception of hazardous areas. Estimators who lack practical on-the-job experience may not always choose the best box type for a given application. The use of such boxes and other ill-adapted materials will cause excessive time to be taken on the job. For example, outlet boxes for use with BX (armored) cable should contain built-in BX clamps; many times boxes will be ordered with knockouts only. This latter case requires the use of additional BX connectors which may only take a few additional seconds per connection, but when these are added up over the period of a large project, much additional time is wasted. Since the cost of labor is an expensive item of cost, any excessive labor required will more than offset any savings gained from the use of poor quality or inadequate materials.

Excessive labor is often consumed due to obstructions that enter conduit installations during the general construction work. Most of these obstructions can be avoided by plugging all conduit

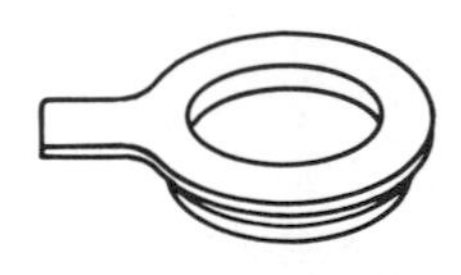
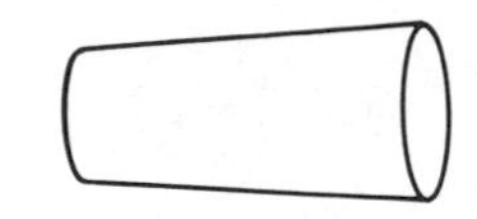
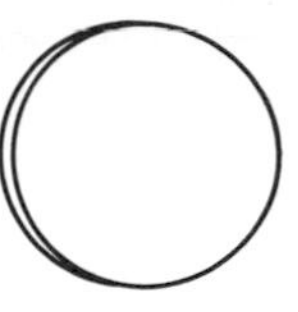

Figure 5.2. Most obstructions that enter conduit installations can be prevented by plugging the conduit terminals with push pennies, wooden plugs, or similar means of protection.

terminals with capped bushings or some similar means of protection. Care should also be taken to thoroughly tighten all fittings, couplings, and so on. In the case of outlet boxes installed in poured concrete structures, painting the inside of the boxes with grease and using capped bushings will greatly aid in keeping concrete out of the raceway system. If any concrete should enter the box, the grease will prevent it from adhering to the metal box.

The electrical contractor should strive to assign electricians to the types of work in which they are experienced or have aptitude to cut down on excessive labor caused by lack of experience on a certain type of work. Of course, this is not always possible, especially in times of manpower shortage when the contractor must use any and all personnel sent out by the union's business manager or obtained otherwise. This factor, however, may be completely or partially overcome by providing proper instructions (supplemental working drawings and specifications) or guidance on the job by an experienced superintendent or foreman.

Errors in locating conduit stub-ups in concrete slabs seem to be one of the greatest difficulties facing the contractor during the rough-in stage of the construction. Careful measurements and rechecking can limit these mistakes to a minimum.

Errors in measurement of large feeder conductors for conduit pulls can be a very costly mistake. When the conductors are cut too short, not only are materials wasted, but the error also causes extra labor in remeasuring, cutting, and installing new conductors. Pulling large sizes of feeder conductors upward in the case of vertical runs when the pull would have been just as easy to set up for a downward pull.

Labor is also wasted by installing individual runs of conduit one at a time in the case of multiple or parallel runs of conduits,

Figure 5.3. Errors in locating conduit stub-ups in concrete slabs seem to be one of the greatest difficulties facing the contractor during the rough-in stage of the construction.

Figure 5.4. Errors in measurement of large feeder conductors for conduit pulls can result in a very costly mistake.

instead of progressively installing all conduits in the run at the same time.

5.2 Planning

Planning is the selection of a plan to achieve a goal. In the case of electrical contractors, their goal is to obtain sufficient work that will enable them to keep the organization intact by providing them with sufficient funds to meet all operating costs and to make a reasonable profit. This goal can only be obtained by proper planning and good job management once the project has been obtained and is under construction.

The need of preparing job progress schedules prior to starting any work was pointed out in Chapter 3. However, once the contractor and crew begins actual work on the job site, further planning and redistribution may be necessary in order to arrive at a more accurate sequence distribution of the labor requirements. Job planning at this stage of the construction involves setting up a list of work operations approximating the sequence of the installations, and redistributing the man-hours as necessary. The result provides a basis for determining the number of electricians required during any given period of the job's progress.

Once the estimated labor has been resegregated as described previously, the estimated man-hours for each sequence section of work operations must be distributed over the period of the job construction during which that particular work will be performed. For example, the labor allowed for all work that would be concealed in the pour of reinforced concrete deck would be distributed by weeks in accordance with a weekly calendar schedule for each period that such work would be performed prior to each pour. Labor allowed for work concealed above suspended ceilings or in hollow partitions would be distributed for the periods that such work would be performed prior to the closing in of the area.

Such planning will help to avoid the consuming of more labor on a particular phase of the project than was allowed in the estimate for that section of the work.

In the case of special materials and equipment not regularly encountered on the conventional electrical contract, it may be necessary to plan the best method of instructing workmen on the installation procedures.

5.3 Organizing

The first step of organizing the work on the job is to hold a conference of at least the supervisory personnel and key workers along with perhaps the estimator who performed the initial estimating of the project. In general, topics such as special installations, general nature of the job, and the need of any special approach in making certain installations should be discussed. Then it should be determined how to best instruct the workmen in the special installation techniques.

It might be advisable to obtain the assistance of outside personnel, like manufacturers' representatives and others more familiar with special job organizational procedures. It should also be determined if the manufacturer's product literature explaining the materials, their application, special installation techniques involved or special job organization outline would be helpful.

The extent of necessary special training should be decided upon and also whether it would be best to hold the training at the job site or at the consultant's place of business. A training schedule should then be drawn up and persons to conduct the training program selected.

Finally, the contractor or his superintendent should proceed with actual arrangements for the training program. Such steps may include scheduling outside assistance, procuring any desirable literature, preparing job organization outlines, procuring demonstration practice materials and tools, setting up the training location, and instructing all concerned regarding their participation in the training program.

5.4 Activating

This phase of the installation deals with promoting a desire of the work team to want to achieve the objectives willingly and in keeping with the planning and organizing efforts. It consists of motivating, directing, and communicating to the people working on the project. In short, activating is getting things done. No amount of planning and organization will get results until the workmen are taught what to do.

A superintendent or job foreman should be able to motivate his people to the highest point of productivity. He must know what kind of workmen are needed to fill the various positions to which he intends to delegate. He must further make sure that the workmen who join his organization are compatible both with himself and those with whom they will work. We realize that this latter statement may not always be possible to fulfill, especially in times of manpower shortages, but it should at least be a goal to reach if at all possible.

One of the most important factors in activating the work on a project is the morale of the workers, and the morale of the workers is directly related to the efficiency with which the job is managed as a whole. If the material and tool servicing is sloppy, incorrect items delivered, and delays are experienced in delivery of materials, the morale drops. Other items that could cause low morale among employees are poor drawings and specifications, poor general supervision, and late payroll deliveries.

In other words, if the workers feel that the contractor is not interested in his own project, allowing some of the problems stated previously to happen, the workers will normally form the same attitude and slack up on production.

If the electrical contractor and his supervisors do their part, the only items left to activating workers is to give directions, instructions, and guidance. Directions tell the workers what, when and where to perform while instructions show the workmen how to perform the work. Job guidance is telling and showing the workmen about a given installation technique when they are not able to perform the work or installation entirely on their own.

5.5 Coordinating

Coordination involves the uniting of the workmen's efforts so that the direct and indirect costs of performing and servicing the project are reduced to the lowest possible minimum. This helps to insure a profitable business operation.

Coordinating work performance and related servicing operations is the chief concern of electrical contractors and their supervisory personnel. The following areas of coordination are of the most importance:

1. Between the estimator who originally estimated the project and other supervisory personnel connected with the particular project.
2. Between general supervisory personnel and those responsible for procurement and servicing the jobs with materials, tools, and work instructions.

3. Between general supervisory personnel and on-the-job supervisory personnel.
4. Between supervisory personnel and the source of manpower and its representatives, including the shop steward in the case of organized labor.
5. Between supervisory personnel and the project owners, architect, and engineers.
6. Between supervisory personnel and the general contractor, other trade contractors, and their related personnel.
7. Between supervisory personnel and electrical workers.
8. Between supervisory personnel and the inspection authorities.
9. Between supervisory personnel and the accounting department with respect to the time records and payroll.
10. Between supervisory personnel and sales personnel, manufacturer's representatives, and so on.

5.6 Controlling

If the planning, organizing, and activating functions are performed as they should be, there will be little need for controlling. However, humans have a tendency to err occasionally, and controlling is therefore important on every project. Ideally, it means recognizing and preventing unwanted happenings from taking place. If this is not possible, the next best is to recognize unwanted happenings at the moment of occurrence so that appropriate action may be taken to correct them.

Control is normally thought of as only a management function. However, proper control begins with the workmen on the job: preventing accidents that would be harmful to workmen, materials, tools, the building, and so on. Foremen must exercise control to see that their crews are kept busy and provided with the proper amount of materials and the correct tools with which to perform the installation. The superintendent should practice control by supporting his foremen and keeping the necessary records current. However, simply to compare performance to the project's estimate, spot trouble areas where variances occur, and similar items is not enough. If nothing is done about correcting these areas, the value of control is meaningless.

6

Branch Circuit Installation Techniques

Most of the material in this chapter deals with branch circuit wiring—that portion of the electrical system covering such items as outlet boxes of all types, all 1-inch and smaller conduit with fittings, and types of cable and building wire up to and including No. 8 AWG. In general, branch circuits include all wiring for lighting, receptacles, small power, and communication systems.

Specific items covered include branch circuit rough-in, surface metal raceway systems, branch circuit wiring, busway, and branch circle cable.

6.1 Branch Circuit Rough-in

An electrical raceway system is designed expressly for holding wires, and in addition to rigid, EMT, or PVC conduit, includes the outlet boxes and other fittings through which the conductors of the system will be installed. For general building construction, rigid or PVC conduit is normally used in and under concrete slabs, while EMT is used for all above-surface installations except where the system will be exposed to severe mechanical injury.

Because of the highly competitive field of electrical contracting, the electrical contractor must be extremely careful about planning the branch circuit installation for a given project. The following items should be given consideration:

1. Buy materials in large quantities where possible—not on speculation, but where materials for several projects that will be completed within the near future can be reasonably predicted.

Figure 6.1. For general building construction, rigid or PVC conduit is normally used in and under concrete slabs.

Figure 6.2. EMT is used for most above-surface installations except where the system will be exposed to severe mechanical injury.

2. Select items of material that will save installation time. For example, a certain type of outlet box may require less time to install than another. If the saving in labor is more than the additional cost of the outlet box, then it will probably pay to use the time-saving outlet box. Also, when it is known that a certain project is going to require many ganged switch boxes, usually some labor can be saved by buying the required boxes from the suppliers rather than having the electricians gang individual boxes on the job.
3. Keep all tools in good working condition. An electrician using a drill motor with a defective switch cannot do his best work if he has to operate the switch several times to make contact each time he uses it.
4. Look into the possibility of building "custom" tools to help reduce labor. Many such tools have been invented by experienced electricians, but were never manufactured for one reason or another. One such device consists of a stick with four markers located at various intervals along the stick. The bottom stick is used to mark the location of conventional receptacle, television, and telephone outlet boxes; the second from the bottom locates outlet boxes above countertops; the next one marks switch boxes; and the top marker is used to locate wall-mounted lighting fixture outlets.

   One contractor made a platform with swivel wheels on which a small seat enclosed space for several dozen duplex receptacles. Using this platform along with a ratchet screwdriver, his electricians were able to install receptacles at a rate nearly 40% faster than by conventional means. An automobile mechanic's roller seat with a built-in tool tray can be purchased for less than $10 and will work just as well.
5. Don't purchase unnecessary tools, but don't be stingy on tools that can earn your firm additional profits. For example, a small cable cutter purchased at under $25 will pay for itself on only a few projects in cutting cable that normally would have been cut with a hacksaw.

6.1.1 Rigid Nonmetallic Conduit

Rigid nonmetallic conduit and fitting (PVC) electrical conduit) may be used where the potential is 600 volts or less in direct earth burial not less than 18 inches below the surface. If less than 18 inches, the PVC conduit must be encased in not less than 2 inches of concrete.

PVC conduit may further be used in walls, floors, ceilings, cinder fill, and in damp and dry locations except in certain hazardous locations, for support of fixtures or other equipment, and where subject to physical damage.

PVC conduit can be cut easily at the job site without special tools. Sizes ½ inch through 1½ inches can be cut with a fine-tooth handsaw. For sizes 2 through 6 inches, a miter box or similar saw guide should be used to keep the material steady and assure a square cut. In order to assure satisfactory joining, care should be taken not to distort the end of the conduit when cutting.

After cutting, deburr the pipe ends and wipe clean of dust, dirt, and plastic shavings. Deburring is accomplished easily with a pocket knife or file.

One of the important advantages of PVC conduit, in comparison with other rigid conduit materials, is the ease and speed with which solvent cemented joints can be made. The following steps are required for a proper joint.

1. Conduit should be wiped clean and dry.
2. Apply a full even coat of PVC cement (use natural bristle brush or spray cement) to the end of the conduit. Cement should cover the area that will be inserted in the socket.
3. Push conduit and fitting firmly together with slight twisting action until it bottoms and then rotate the conduit in the fitting (about a half turn) to distribute the cement evenly. Avoid cement buildup in ID of conduit. The cementing and joining operation should not exceed more than 20 seconds. Let dry for approximately 10 minutes.

When the proper amount of cement has been applied, a bead of cement will form at the joint. Wipe joint with brush to remove excess cement. Joints should not be disturbed for ten minutes at room temperatures. The chart in Figure 6.3 lists the average amount of cement applicable to the type of fitting.

Most manufacturers offer various radius bends in a number of segments. Where special bends are required, PVC conduits are easy to form on the job. The descriptions in Figures 6.4–6.8 give examples of commonly used bends.

Very little practice is required to master these techniques. Almost no calculations are required in laying out field bends. Conduit section to be bent must be heated evenly over the entire length of the curve. Most manufacturers offer electric heaters designed specifically for the purpose in sizes to accommodate all conduit diameters. These devices employ infrared heat energy which is most quickly absorbed by the conduit. Small sizes are ready to bend after a few seconds in the "hotbox." Larger diame-

Size Fitting	PINT* No. of Joints	QUART No. of Joints
½″	350	700
¾″	200	400
1″	150	300
1¼″	110	220
1½″	80	160
2″	45	90
3″	35	70
3½″	30	60
4″	25	50
5″		25
6″		15

Figure 6.3. Average amount of cement applicable to various sizes of fittings.

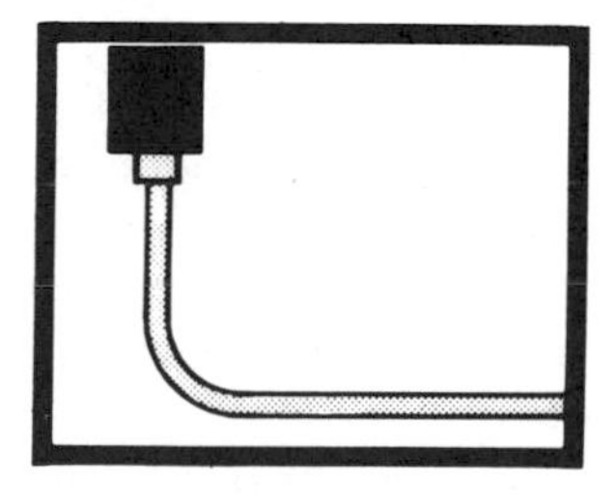

Figure 6.4. Stub: Usually formed near the end of a conduit section with a rise of 12–18 inches to the location of an outlet box or other enclosure. Quick and easy to make.

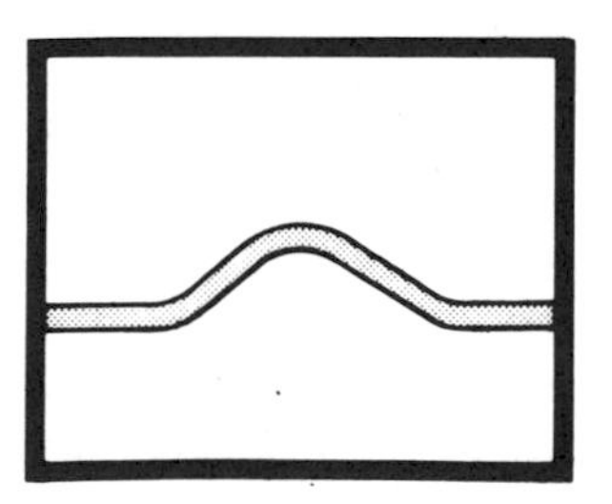

Figure 6.5. Saddle: Used to ease the conduit over an obstruction such as a pipe or a beam. PVC's smooth interior assures easy wire pulling even in tight bends.

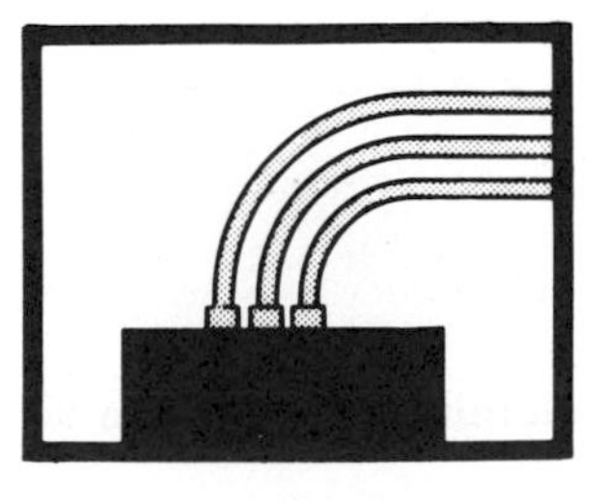

Figure 6.6. Concentrics: A series of conduits turning together around a common center. Each requires a different radius. Easily done with PVC conduit.

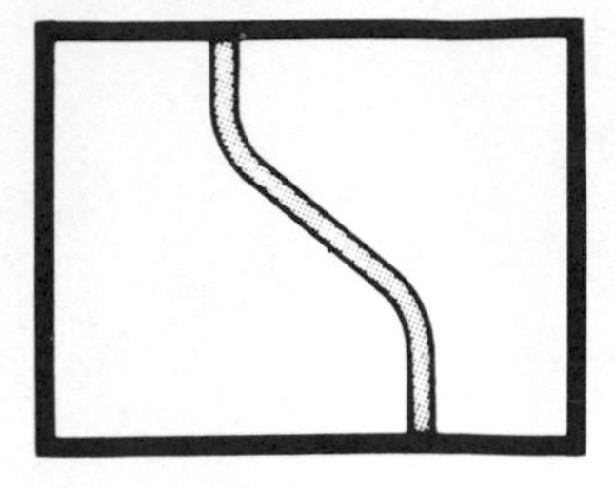

Figure 6.7. Offset: Change in the line of a conduit run to avoid an obstruction or to meet an opening in an enclosure. Can be "tailored to fit" with PVC conduit.

ters require two or three minutes depending on conditions. Other methods of heating PVC conduit for bending include heating blankets and hot air blowers. Immersion in hot liquids (about 275°F) is also satisfactory. The use of torches or other flame-type devices is not recommended. PVC conduit exposed to excessively high temperatures may take on a brownish color. Sections showing evidence of such scorching should be discarded.

If a number of identical bends are required a jig can be helpful (see Figure 6.9). A simple jig can be made by sawing a sheet of plywood to match the desired bend. Nail to a second sheet of plywood. The heated conduit section is placed in the jig, sponged with water to cool, and it's ready to install. Care should be taken to fully maintain I.D. of the conduit when handling.

If only a few bends are needed, scribe a chalk line on the floor or workbench. Then match the heated conduit to the chalk line and cool. The conduit must be held in the desired position until relatively cool since the PVC material will tend to go back to its original shape. Templates are also available for any desired bend that comply with the NEC requirements.

Another method is to take the heated conduit section to the point of installation and form it to fit the actual installation with the hands (see Figure 6.10). Then wipe a wet rag over the bend (Figure 6.11) to cool it. This method is especially effective in making "blind" bends or compound bends.

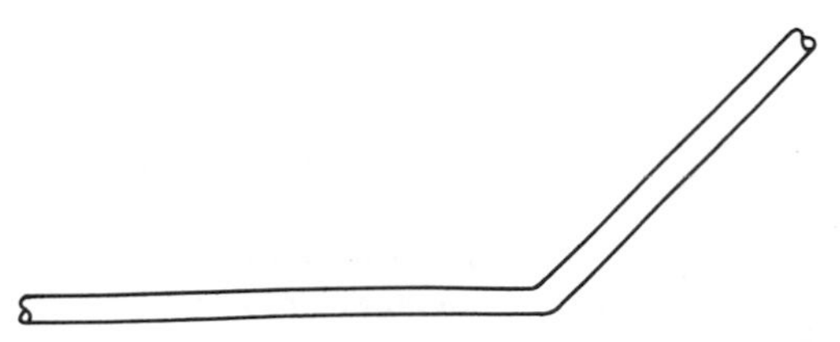

Figure 6.8. Kick: Minor change in direction of a conduit run. Often can be done "cold" in slab work.

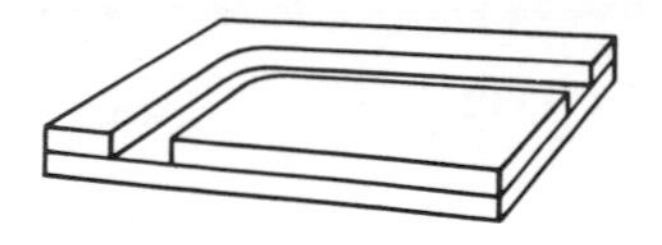

Figure 6.9. If a number of identical bends are required, a jig made by sawing a sheet of plywood to match the desired bend can be used.

Bends in small-diameter PVC conduits (½–1½ inches) require no filling for code approved radii. In larger sizes the interior must be supported to prevent collapse for other than minor bends. Use a flexible spring or air pressure.

Place air-tight plugs (Figure 6.12) in each end of the conduit section before heating. The retained air will expand during the heating process and hold the conduit open during the bending. Do not remove the plugs until the conduit is cooled.

In applications where the conduit installation is subject to constantly changing temperatures and the runs are long, precautions should be taken to allow for expansion and contraction of PVC conduit.

When expansion and contraction are factors, an "O" ring expansion coupling should be installed near the fixed end of the run, or fixture, to take up any expansion or contraction that may occur. Confirm the expansion and contraction length available in these fittings as it may vary by manufacturer. The chart in Figure 6.13 indicates what expansion can be expected at various temperature levels. The coefficient of linear expansion of PVC conduit is 0.0034 inch/10 ft/°F.

Expansion couplings are seldom required in underground or slab applications. Expansion and contraction may generally be controlled by bowing the conduit slightly or immediate burial.

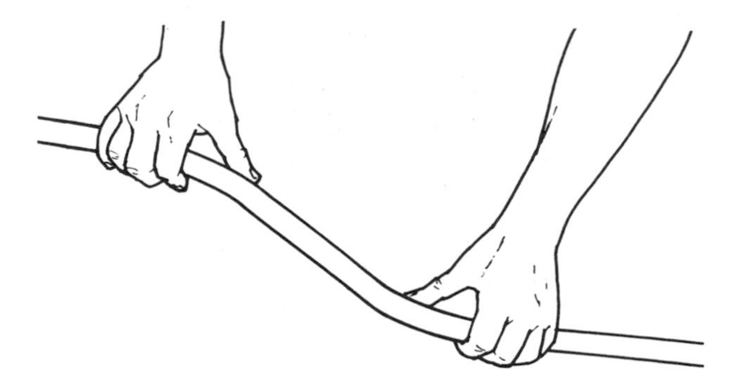

Figure 6.10. Another method is to take the heated conduit section to the point of installation and form it to fit the actual installation with the hands.

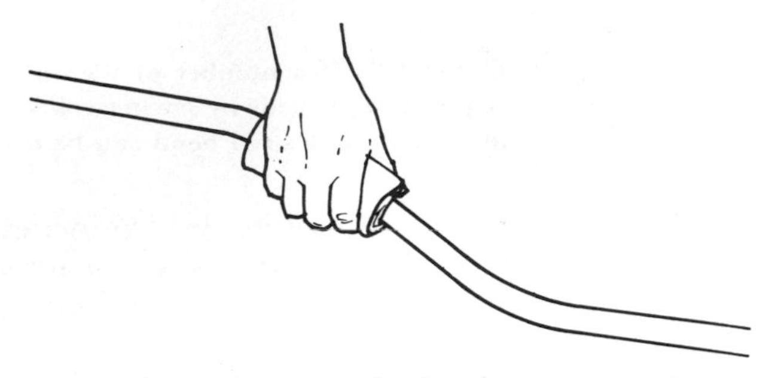

Figure 6.11. After the bend is formed, wipe a wet rag over the bend to cool it.

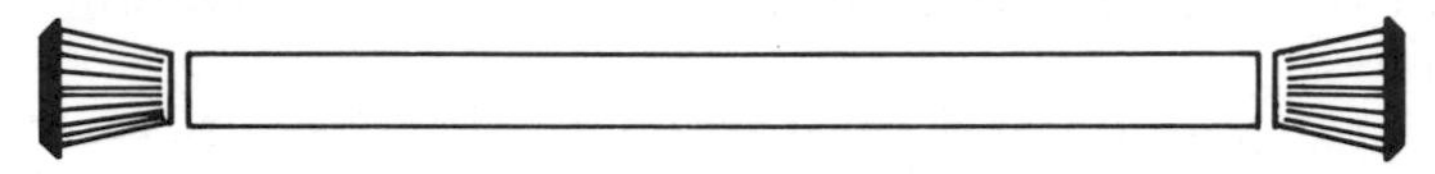

Figure 6.12. Air-tight plug should be placed in each of the conduit sections before heating.

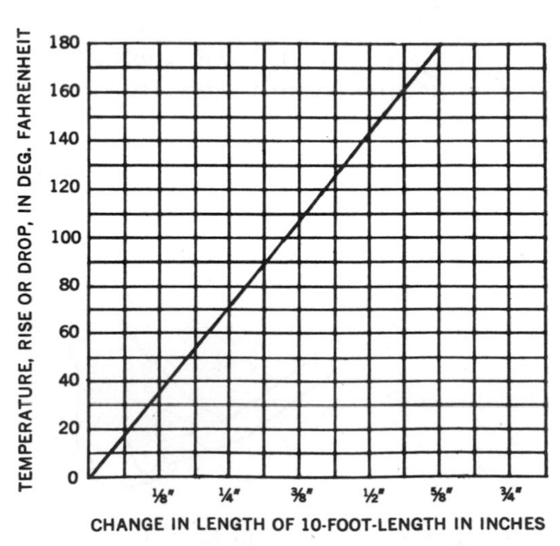

Figure 6.13. The expansion to be expected at various temperature levels.

After the conduit is buried expansion and contraction ceases to be a factor. Care should be taken, however, in constructing a buried installation. If the conduit should be left exposed for an extended period of time during widely variable temperature conditions, allowance should be made for expansion and contraction.

In above-ground installations, care should be taken to provide proper support of PVC conduit due to its semirigidity. This is particularly important at high temperatures. Distance between supports should be based on temperatures encountered at the specific installation. The chart in Figure 6.14 clearly outlines at what intervals support is required for PVC conduit at various temperature levels.

6.1.2 Rigid Metal Conduit

Galvanized rigid metal conduit may be used under all atmospheric conditions and occupancies. However, unless corrosion protection is provided, rigid metal conduit and fittings should not be installed in areas subject to severe corrosive influences.

During the installation of rigid conduit, it has been the practice to use special conduit cutters to cut the conduit. These cutters

RECOMMENDED CONDUIT SUPPORT SPACING							
	Maximum Temperature in Degrees Fahrenheit						
Nominal Size	20° ft.	60° ft.	80° ft.	100° ft.	120° ft.	140° ft.	160° ft.
½"	4	4	4	4	2½	2½	2
¾"	4	4	4	4	2½	2½	2
1"	5	5	5	5	3	2½	2½
1¼"	5	5	5	5	3	3	2½
1½"	5	5	5	5	3½	3	2½
2"	5	5	5	5	3½	3	2½
2½"	6	6	6	6	4	3½	3
3"	6	6	6	6	4	3½	3
3½"	7	7	7	6	4	3½	3½
4"	7	7	7	7	4½	4	3½
5"	7	7	7	7	4½	4	3½
6"	8	8	8	8	5	4½	4

When so marked EPC-40 PVC rigid conduit may be used with 90°C rated cable.

Figure 6.14. Spacing of conduit supports at various temperature levels.

normally leave a large burr, and often a definite hump, inside the conduit requiring additional time to remove the burr. A better method would be to use a lightweight portable electric hack saw using blades with 18 teeth per inch.

Conduit cuts should be made square and the inside edge of the cut adequately reamed to remove any burr or sharp edge which might damage the insulation of the conductors when pulled in later. Lengths of conduit should be accurately measured before they are cut as recutting will obviously result in lost time.

Basically, the contractor should provide a vice stand for each crew that will securely hold the conduit and not shift about as the cut is being made. A power hack saw of either the blade or band type should be provided as well as a sufficient supply of hack saw blades to be used in the electrician's hack saw frames. The cut should be made entirely through the conduit, not broken off the last fraction of an inch. While the hack saw may be used to cut smaller sizes of conduit by hand, the larger sizes should not be cut by hand except in emergencies. Not only will hand cutting of the large sizes take up too much time, but it is also extremely difficult to cut such large sizes of conduit square.

In addition to the length needed for the piece of conduit for a given run, allow an additional ⅜ inch on smaller sizes of conduit for the wall of the box and the bushing. Because larger sizes of conduit are usually connected to heavier boxes, allow approximately ½ inch. If additional locknuts are needed in a run, a ⅛-inch allowance for each locknut will be sufficient. Where conduit bodies are used, include the length of the threaded hub in the measurements.

The usual practice for threading the smaller sizes of rigid conduit is to use a pipe vice in conjunction with a die stock with proper size guides and sharp cutting dies properly adjusted and securely held in the stock. Clean, sharp threads can be cut only when the conduit is well lubricated; use a good lubricant and plenty of it.

Conduits should always be cut with a full thread. To accomplish this, run the die up on the conduit until the conduit just about comes through the die for about one full thread. This gives a thread length that is adequate for most purposes. However, don't overdo it; if the thread is too long, that portion which does not fit into the coupling will corrode because threading removes the protective covering.

Clean, sharply cut threads also make a better continuous ground and save much grief and unnecessary labor. It takes a little extra time to make certain that threads are properly made, but a little extra time spent at the beginning of a job may save much time later on.

On projects where a considerable number of relatively short sections of conduit will be required for nipples, considerable threading time can be saved by periodically gathering up the short lengths of conduit resulting from previous cuts and reaming and running a thread on one end with a power threader and redistributing these lengths to the installation points about the job. This procedure eliminates one-hand threading operation in many instances.

When threadless couplings and connectors are used with rigid conduit, they should be made tight. Where the conduit is to be buried in concrete, the couplings and connectors must be of the concrete type, and where used in wet locations, they must be of the raintight type.

The installation of rigid conduit for branch circuit raceways often requires many changes of direction in the runs ranging from simple offsets to complicated angular offsets, saddles, and so on.

In bending elbows, care should be taken to comply with the National Electrical Code. In general, the Code states that an elbow or 90° bend must have a minimum radius of six times the inside diameter of the conduit. Therefore, the radius of 2-inch conduit must have a radius of at least 12 inches, a 3-inch conduit 18 inches, and so forth.

Bends are normally made in the smaller sizes of conduit by hand with the use of conduit hickeys or benders. In some cases, where many bends of the same type must be made, hand roller "Chicago benders" or hydraulic benders are used to simplify making the bends to certain dimensions.

Occasionally, rigid conduit will have to be rebent after it is installed. In such cases, the rebending must be done carefully so that the conduit does not break. Most often, these rebends will have to be made at stub-ups—conduits emerging through concrete floors. To rebend conduit, the concrete should be chipped away for a few inches around the conduit and then the conduit should be warmed with a propane torch. It can then be bent into the required shape without further trouble.

6.1.3 Electrical Metallic Tubing

Electrical metallic tubing may be used for both exposed and concealed work except where the tubing will be subjected to severe physical damage, in cinder concrete unless the tubing is at least 18 inches under the fill.

The tubing should be cut with a hand or power hack saw using blades with 32 teeth per inch, after which the cut ends should be reamed to remove all rough edges. Threadless couplings and connectors used must be made tight and the proper type should be used for the situation; that is, concrete-tight types should be used when the tubing is buried in concrete, and raintight type used when installed in a wet location. Supports must be provided, when installed above grade, at least every 10 feet and within 3 feet of each outlet box or other termination point.

Bends are made in EMT much the same as for rigid conduit. However, roll-type benders are used exclusively for EMT. This type of bender has high supporting sidewalls to prevent flattening or kinking of the tubing and a long arc that permits the making of 90° bends (or any lesser bends) in a single sweep.

Many time-saving tools and devices have appeared on the market during the past years to facilitate the installation of electrical metallic tubing. Table hydraulic speed benders, for example, makes 90° bends or offsets in 5–10 seconds. Shoes are available for ½-inch through 1¼-inch EMT. Mechanical benders are also available for sizes through 2-inch EMT.

An automatic "kinker" will eliminate the need for offset connectors wherever ¾-inch or ½-inch EMT is used. The end of a piece of EMT is inserted in the blocks of the device and one push of the handle (about 2 seconds duration) makes a perfect offset. Every bend is identical, which eliminates lost time refitting or cutting and trying.

EMT hand benders with built-in degree indicators lets the operator make accurate bends from 15° to 90° faster since in-between measurements are eliminated.

6.2 Surface Metal Raceway

Surface metal raceway is one of the exposed wiring systems that is quite extensively used in existing buildings where new wiring or extensions to the old are to be installed. Although it does not afford such rugged and safe protection to the conductors as rigid conduit or EMT, it is a very economical and quite dependable system when used under conditions for which it was designed. The main advantage of surface metal raceway is its neat appearance where wiring must be run on room surfaces in finished areas.

Surface metal raceways may be installed in dry locations except where subject to severe physical damage or where the voltage is 300 volts or more between conductors (unless the thickness of the metal is not less than 0.040 inch). Furthermore, surface metal raceways should not be used in areas that are subjected to corrosive vapors, in hoistways, nor in any hazardous location. In most cases, this system should not be used for concealed wiring.

Various types of fittings for couplings, corner turns, elbows, outlets, and so on are provided to fit these raceways. Figure 6.15 shows a number of the most common fittings in use.

Many of the rules for other wiring systems also apply to surface metal raceways; that is, the system must be continuous from outlet to outlet, it must be grounded, and all wires of the circuit must be contained in one raceway and so on.

In planning a surface metal raceway system, the electrical contractor should make certain that all materials are provided before the installation is begun. One missing fitting could hold up the entire project.

Proper tools should also be provided to make the installation easier. For example, a surface metal raceway bending tool will enable electricians to bend certain sizes of the raceway like rigid conduit or EMT. With such a tool, many of the time-consuming fittings can be eliminated as the tool can make 90° bends, saddles, offsets, back-to-back bends, and so on. Some bending tips for the Wiremold Benfield Bender follows:

1. *Bending with tool in air:* Apply hand pressure as close to the tool as possible. Keep pressure close to the groove for smoother bends and greater accuracy.
2. *Bending on floor:* Work on hard surfaces. Avoid soft sand or deep pile carpets.
3. *Degree scale:* One side of the tool (closed hook side) is calibrated for 500 Wiremold. The opposite, open hook side, is scaled for 700.
4. *Zero (0°) degree line:* The zero degree line in bottom of groove adjacent to the hook is the point of beginning of the bend.

For Wiremold 200 only

200F FLEXIBLE SECTION
18" l. overall.

201 COUPLING
1½" l.

202 BUSHING
Protects wires from abrasion at open end of 200.

203 SUPPORTING CLIP
2½" l.

205 STRAP
(One-hole)

206 CONNECTION COVER
For covering gaps in raceway.

211 90° FLAT ELBOW
Base, each leg 1¼" l. without tongue.

211LH INTERNAL TWISTED ELBOW
Base, each leg 2" l. without tongue.

211RH INTERNAL TWISTED ELBOW
Base, each leg 2" l. without tongue.

214 PULL BOX
5" l.

For Wiremold 500, 700, 5700

Because of space limitation, the fittings included on these pages are only briefly described. More complete information is available on request.

5700F FLEXIBLE SECTION
18" l. overall.

5700WC WIRE CLIP

5701 COUPLING

5701A TONGUE ADAPTER

57700 TRANSITION COUPLING
For direct connection of 5700 and 700.

502 BUSHING (500)
702 (700, 5700)
Protects wires from abrasion at open end of raceway.

504 STRAP (500)
704 (700, 5700)
(One- or two-hole)

CONNECTION COVER
506 (500)
706 (700, 5700)
For covering gaps in raceway.

45° FLAT ELBOW
512 (500)
712 (700, 5700)
For diagonal turns on same surface.

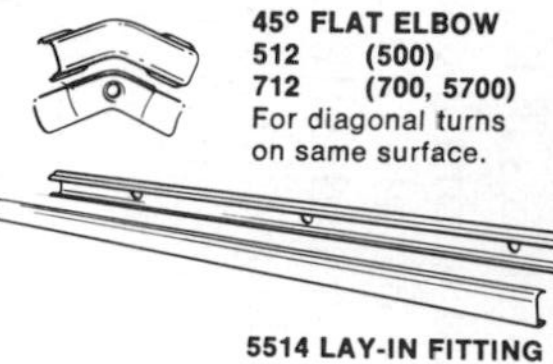

5514 LAY-IN FITTING

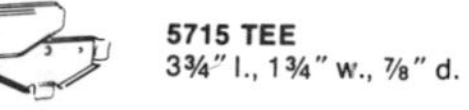

5715 TEE
3¾" l., 1¾" w., ⅞" d.

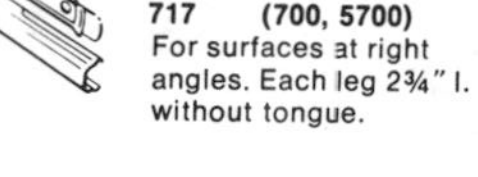

INTERNAL ELBOW
517 (500)
717 (700, 5700)
For surfaces at right angles. Each leg 2¾" l. without tongue.

5717A INTERNAL PULL ELBOW
Makes pulling of wires easy. Base, each leg 5½" l. without tongue.

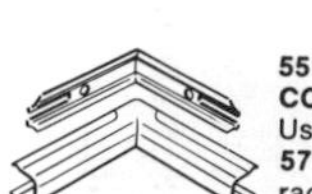

5517C INTERNAL CORNER COUPLING (500)
Use with 5514.
5717C Use with 5700 raceway.

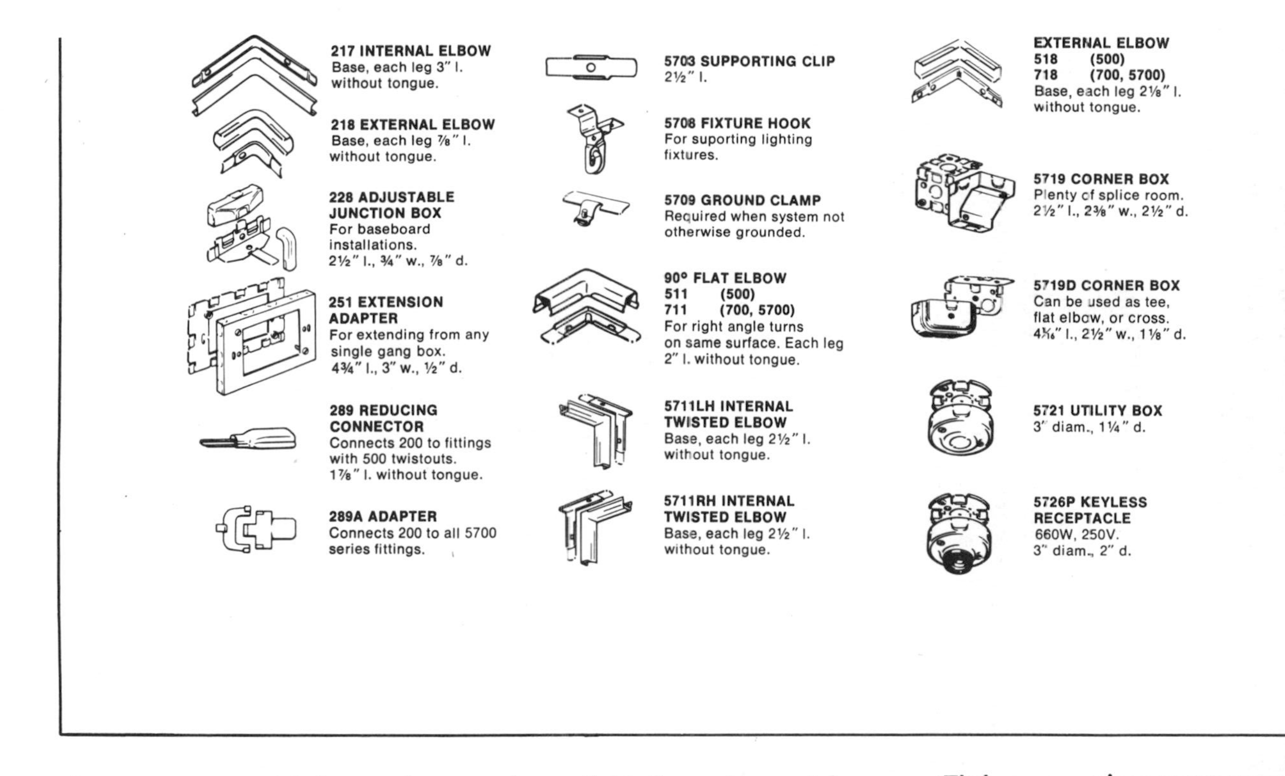

Figure 6.15. Several fittings and accessories available for surface metal raceway. Fittings: one-piece raceways and two-piece 5700. (Courtesy of The Wiremold Co.)

5. *Rim notches:* The rim notches closest to the hook indicate exact center of a 45° bend. Numerals "500" and "700" tell the operator which notch to use for which size Wiremold.

6.3 Branch Circuit Wiring

In most cases, the installation of branch circuit wires is merely routine. However, there are certain practices that can reduce labor and materials to the extent that such practices should at least be given careful consideration. The use of modern equipment, such as vacuum fish tape systems, is another way to reduce labor during this phase of the wiring installation.The proper size and length of fish tape, as well as the type, should be one of the first considerations. For example, if most of the runs between outlets are only 20 feet or less, a short fish tape of, say, 25 feet will easily handle the job and will not have the weight and bulk of a larger one. When longer runs are encountered, the required length of fish tape should be enclosed in one of the metal or plastic fish tape reels. This way the fish tape can be rewound on the reel as the pull is being made to avoid having an excessive length of tape lying around on the floor.

When several bends are present in the raceway system, the insertion of the fish tape may be made easier by using flexible fish tape leaders on the end of the fish tape.

The combination blower and vacuum fish tape systems are ideal for use on long runs and can save much time. Basically, the system consists of a tank and air pump with accessories. An electrician can vacuum (Figure 6.16) or blow (Figure 6.17) a line or tape in any size conduit from ½ through 4 inches, or even 5- and 6-inch conduits with optional accessories.

After the fish tape is inserted in the raceway system, the wires must be firmly attached by some approved means. On short runs, where only a few conductors are involved, all that is necessary is to strip the insulation from the ends of the wires, bend these ends around the hook in the fish tape, and securely tape them in place. Where several wires are to be pulled together over long and difficult (one with several bends) conduit runs, the wires should be staggered and the fish tape securely taped at the point of attachment so that the overall diameter is not increased any more than is absolutely necessary. Staggering is done by attaching one wire to

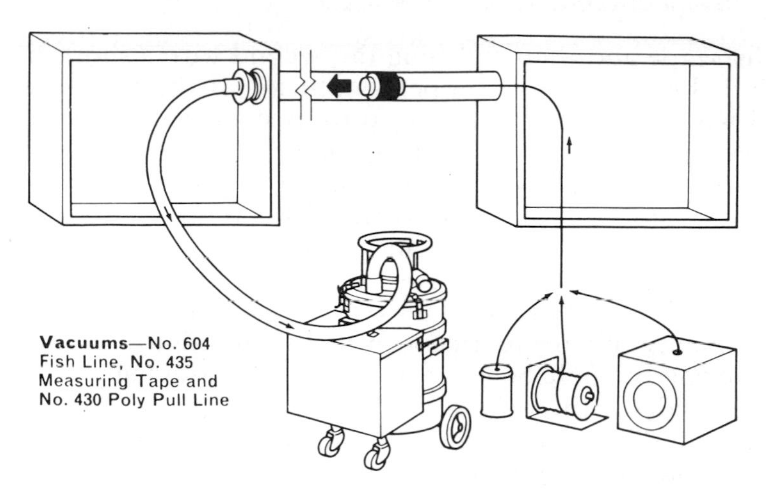

Vacuums—No. 604
Fish Line, No. 435
Measuring Tape and
No. 430 Poly Pull Line

Figure 6.16. The combination blower and vacuum fish tape system can be used to vacuum a pull line in any size conduit.

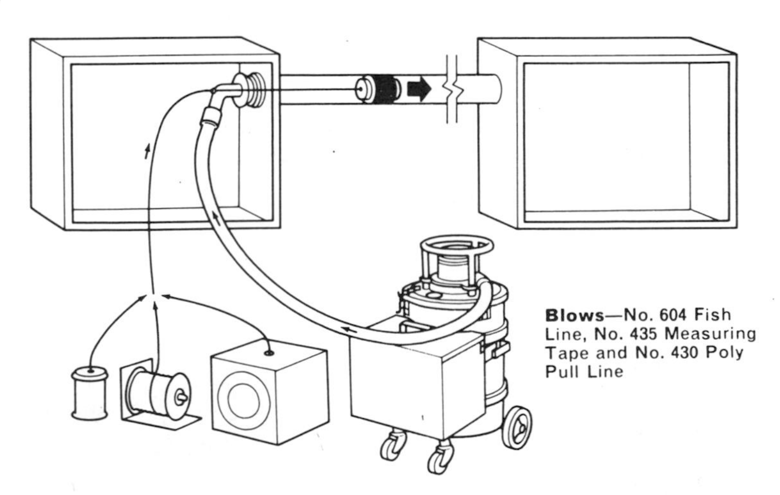

Blows—No. 604 Fish
Line, No. 435 Measuring
Tape and No. 430 Poly
Pull Line

Figure 6.17. The fish tape system may also be used to blow a line through conduit.

the fish tape and then attaching the second wire a short distance behind this to the bare copper conductor of the first wire. The third wire, in turn, is attached to the second wire and so forth.

Basket grips (Figure 6.18) are available in many sizes, for almost any size and combination of conductors. They are designed to hold the conductors firmly to the fish tape and can save the electrical workers much time and trouble that would be required when taping wires.

In all but very short runs, the wires should be lubricated with a good quality of wire lubricant prior to attempting the pull, and also during the pull. Some of this lubricant should also be applied to the inside of the conduit itself.

When pulling wire from coils in boxes, the wire should be pulled from the center of the coil to prevent excessive tangling and kinking. If several conductors are to be fed into a single raceway system, they should be kept parallel and straight, free from kinks and bends. Wires that are permitted to cross each other during the pulling operation form a bulge that makes pulling difficult, especially around bends.

Wire dispensers are great aids in helping to keep the conductors straight and to facilitate the pulling of conductors. One type is designed to accept up to 6 boxes of wire from size No. 16 to No. 10 AWG. A feeding eye in the top of the dispenser helps eliminate kinks and tangles while large 10-inch semipneumatic tires make it convenient to move the entire apparatus (including the boxes of wire) to various locations.

Other types of dispensers are designed for spools or wires. These compact spool wire dispensers handle up to 10 spools of wire from No. 22 to No. 10 AWG. The spindles are designed to permit easy payout of the wire with automatic braking. Guide

Figure 6.18. Basket grips are available in many sizes, for almost any size and combination of conductors.

heads are also provided to gather and straighten the wires for easy feeding, eliminating kinks or tangles. One type even has a guide head that rotates 360° and can be raised or lowered approximately 30° for pulling wires into a conduit system from any angle.

6.4 Branch Circuit Cable

Nonmetallic-sheathed (type NM) and metal clad (type AC) cables are very popular for use in residential and small commercial wiring systems. In general, both types of cable may be used for both exposed and concealed work in normally dry locations. They may be installed or fished in air voids in masonry block or tile walls where such walls are not exposed or subject to excessive moisture or dampness. Type NM cable shall not be installed where exposed to corrosive fumes or vapors; nor shall it be embedded in masonry, concrete, fill or plaster; nor run in shallow chase in masonry or concrete and covered with plaster or similar finish.

Type NM cable must not be used as a service-entrance cable, in commercial garages, in theaters and assembly halls, in motion-

Figure 6.19. Type NM cable is very popular for use in residential and small commercial wiring systems.

picture studios, in storage battery rooms, in hoistways, in an hazardous location, or embedded in poured cement, concrete, or aggregate. Type AC cable has the same restrictions.

For use in wood structures, holes are bored through wood studs and joists first, and then the cable is pulled through these holes to the various outlets. Normally, the holes give sufficient support providing they are not over 4 feet on center. Where no stud or joist support is available, staples or some similar support is required for the cable. The supports must not exceed 4½ feet and must also be supported within 12 inches of each outlet box or other termination point.

Proper tools facilitate the running of branch circuit cables and include such items as sheathing strippers for stripping the NM cable; hack saw for cutting and stripping type AC (BX) cable, a carpenter's apron for holding staples, crimp connectors, and wire nuts.

This chapter has covered many of the detailed relations of electrical wiring methods as they relate to electrical building construction, as well as many related installation techniques.

While the items covered are considered to be good practice at the present time, improved installation practices are certain to develop in the future which will contribute to a greater efficiency in providing electrical installations.

Electrical contractors have a responsibility to themselves, their employees and the public to continually seek improved ways to provide such installations as far as quality and price are concerned.

7

Service and Feeder Wiring

Items normally considered as service and feeder wiring consist of all 1¼-inch and larger raceway systems (including fittings, hangers, etc.), panelboards, safety switches, switchboards, and all wire and cable, No. 6 AWG and larger. Busway systems, grounding, and similar related items fall under this phase of the electrical construction.

In most cases, the service entrance equipment is located at one point in the building either at a central location to keep the length of the feeders and subfeeders at a practical minimum length, or at a point close to where the service entrance conductors enter the building.

Service entrance equipment provides overcurrent protection and a means of disconnecting the feeder and service entrance conductors. Feeders provide a path for electric current from the main service entrance equipment to subpanels, dry transformers, HVAC equipment, and other heavy electrical equipment and devices.

This chapter deals with the best methods of handling the installation of service and feeder wiring, as well as related items.

7.1 Service and Feeder Rough-in

As soon as the building lines have been established on any given project, the electrical contractor should make arrangements to provide and install all necessary conduit sleeves before the footings are poured. This is also the time to orient plans for installing all heavy electrical equipment in relation to fixed and identical building points.

When these points have been established, the electrical contractor can positively fix the location of main distribution service equipment, panelboards, transformers, and other equipment requiring large conduit sizes and conductors. Once it has been established that service and feeder conduits will not interfere with the general building construction, these raceways should be installed. The reasons for this early installation are numerous, but some of the advantages include:

1. Service and feeder raceways run under the concrete slab usually require less materials as well as less labor.
2. Positioning the service and feeder raceways prior to the first pouring enables the electrical contractor to make any necessary diggings to bury the large size conduits; this would not be possible once the first pour has been made.

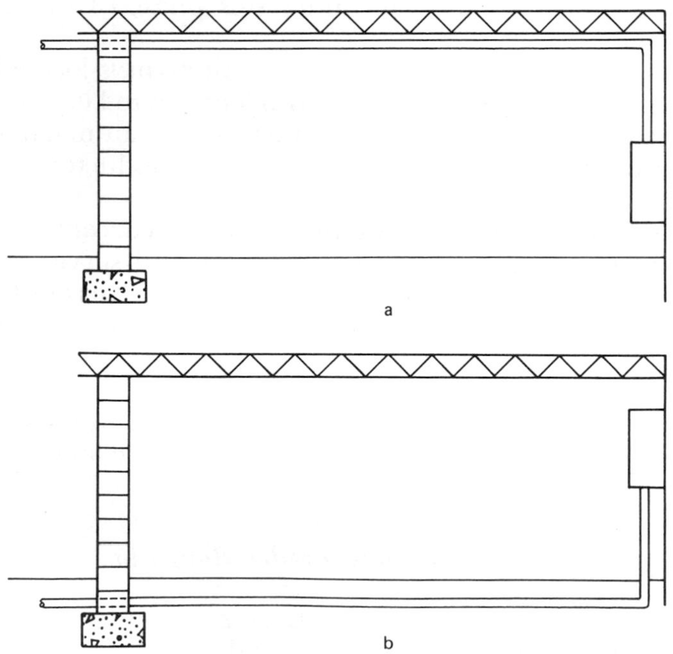

Figure 7.1. (a) Sectional view of building with panelboard feeder installed overhead. (b) Sectional view of building with panelboard feeder installed underground.

A further explanation of item No. 1 preceding is in order, as many contractors may disagree with this statement. Their concensus is that if a panelboard was located in a building such as the one in Figure 7.1a, what would be the difference in material and labor if the circuit was run underground as is shown in Figure 7.1b? In fact, for the larger size conduits, some trenching may be required for the conduit in the underground installation.

While it is true that the same amount of conduit and wire is involved in both installations (Figures 7.1a and b), and the trenching may require some extra time, the installation could normally be done cheaper by running the conduit underground. First of all, the trenching will be made in fill dirt most of the time and one man with a shovel can quickly open such ground. Therefore, the extra time required for the trenching is not going to be too great. In fact, a trench used for plumbing pipe or other utility may possibly be used.

Secondly, the installation of the conduit at ground level, requiring no ladders or scaffolding, can be made in much quicker time than overhead; for a 100-foot run of 4-inch conduit, about 7

Figure 7.1c. Conduit installations overhead require the use of ladders or scaffolding which consume more time than if the installation were at ground level.

man-hours less! The installation at ground level also eliminates the need for hangers and similar miscellaneous items.

Finally, if the installation of the feeder is delayed so that it must be run overhead, chances are other trades, equipment and workers, are going to be in the way. This would cause even more man-hours to complete the installation.

When installing parallel runs of conduit, always make certain that enough space is allowed between each run for locknuts, bushings, and couplings. Also the parallel runs should be installed simultaneously rather than progressively installing one single raceway at a time.

Whenever possible, the entire raceway system should be completed before pulling in any wire. If the conductors are installed before all the raceway system is finished, it is possible that some wire may be damaged, requiring costly extra labor and materials to correct the damage.

Pull boxes and junction boxes are provided in a raceway system to facilitate the pulling of conductors or to provide a junction point for the connections of conductors or both. In some instances the location and size of pull boxes is designated on the drawings or in the written specifications supplied by the architect or his engineer. However, in most cases, it is the contractor's responsibility to size these boxes correctly according to good practices or the NEC.

The National Electrical Code specifies certain minimum dimensions for a junction box installed in a raceway system utilizing 1 inch or larger conduit containing No. 6 AWG conductors or larger. These rules also apply to pull or junction boxes in cable installations, but instead of using the cable diameter, the minimum trade size conduit for the number and size of conductors in the cable must be used in the calculation.

Figure 7.2 shows a junction box with several conduits entering it. Since 4-inch conduit is the largest size in the group, the

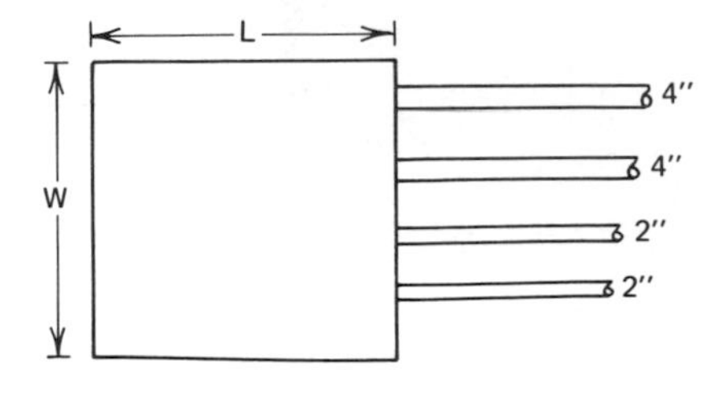

Figure 7.2. Junction box with two 4-inch conduits and two 2-inch conduits entering it.

minimum length required for the box can be determined by the following calculation:

4″ (size of largest conduit) × 8 (as stated in the NEC) = 32″

Therefore, the junction box must be 32 inches in length. The width of the box, however, need only be of sufficient size to enable locknuts and bushings to be installed on all the conduits or connectors entering the enclosure.

Junction or pull boxes in which the conductors are pulled at an angle as shown in Figure 7.3 must have a distance of not less than six times the trade diameter of the largest conduit. The distance must be increased for additional conduit entries by the amount of the sum of the diameter of all other conduits entering the box on the same side; that is, the same wall of the box. The distance between raceway entries enclosing the same conductors must not be less than six times the trade diameter of the largest conduit.

Since the 4-inch conduit is the largest of the lot in this case, $L_1 = 6 \times 4 + (3 + 2) = 29$. L_2 is calculated the same way; therefore, $L_2 = 29$.

The distance $(D) = 6 \times 4$ or 24 inches and this is the minimum distance permitted between conduit entries enclosing the same conductor.

As before, the depth of the box need only be of sufficient size to permit locknuts and bushings to be properly installed.

Regardless of whether junction or pull boxes are specified for a particular project, long runs of wires should not be made in one pull. Pull boxes, installed at convenient intervals relieve much of the strain on the wires and make the pull much easier. Since there is no set rule concerning the distance between pull boxes, the

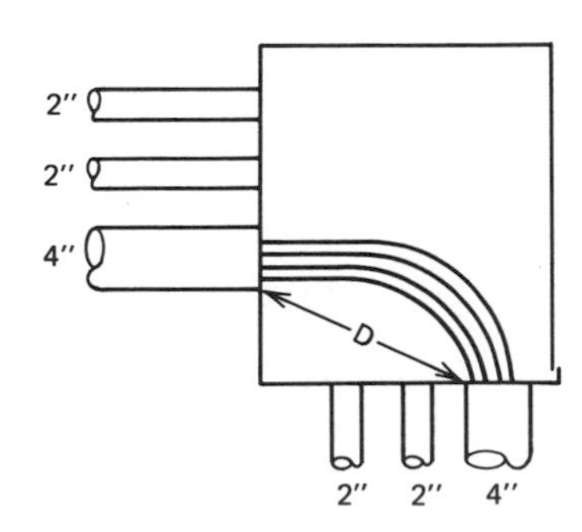

Figure 7.3. Junction box with conduit runs entering at right angles.

electrical contractor or his workmen will have to use good judgment as to where they are required or would be beneficial.

At first glance, the installation of pull boxes in a raceway system may seem to require much extra work, but pull boxes can save a considerable amount of time and hard work when pulling in large size conductors. Properly placed, they eliminate many bends and elbows and do away with the necessity of fishing from both ends of a conduit run.

Pull boxes should always be installed in a location that allows the workmen to work easily and conveniently. In an installation where the conduit run is routed up a corner of a wall and changes direction at the ceiling, a pull box that is installed too high will force the electrician to stand on a ladder when feeding or pulling wires. This situation is just an example; there are several situations to consider when selecting a location for junction boxes. Then they must be securely fastened in place on walls or ceiling or some other adequate support.

7.1.1 Conduit Bends

Although there are not as many conduit bends in service and feeder conduit systems as there are in branch circuit wiring, there are still a few installations where factory ells would take care of all the required bends. The electricians are required to make several complicated offsets, saddles, radius bends, and other bends requiring both skill and calculations.

The National Electrical Code specifies the minimum radius of conduit bends which varies from six to eight times the normal inside diameter of the conduit. The table to follow gives the minimum radius of conduit bends for 1¼ through 6-inch conduit sizes.

Size of Conduit (inches)	Min. Radius of Bend (inches)
1¼	8
1½	10
2″	12
2½	15
3″	18
3½	21
4″	24
5″	30
6″	36

7.1.2 *Grounding*

To make certain that the raceway system provides a continuous equipment ground, all connections between sections of conduit and between the conduits and termination points (panels, wire troughs, junction boxes, etc.) must be tight. This is insured by the use of two locknuts on every termination point even though metal bushings are used.

Where conduits enter a building at the main distribution panel and the grounding is provided at this point, it is a good idea to install bonding bushings on the panel to insure continuity. Electrical disturbances may occur in the wiring system, both inside and out, and many contractors do not like to rely entirely on locknuts and bushings for electrical continuity at this point in the electrical system where a good ground is very important.

After the ground is made, the electrical contractor should check the system with a megger to make certain that the resistance is low enough to qualify under the requirements of the NEC. Where a ground wire is connected to a water pipe, the area around where the ground clamp will be secured should be thoroughly cleaned of all dirt, paint, rust, or other substances that might offer resistance to the flow of electricity.

7.2 *Supports, Hangers, and Fasteners*

The selection of the proper hangers and supports is a very important function in roughing-in of service and feeder raceways. While the electrical estimator will include some hangers and supports in his bid, more often he will allow a lump sum figure for such items and leave the exact selection of these devices to men in the field.

Raceways installed exposed on masonry surfaces are usually held in place with one- or two-hole straps, the straps being securely anchored with lead anchors, toggle bolts, and similar means.

Two or more parallel raceway systems surface-mounted are usually installed on special metal channels and are then held in place by matching pipe straps. This method seems to be the most popular since it not only is the fastest way to install banks of conduit, but also gives a very neat appearance. Figure 7.4 shows several applications of metal support channels and single-bolt pipe straps.

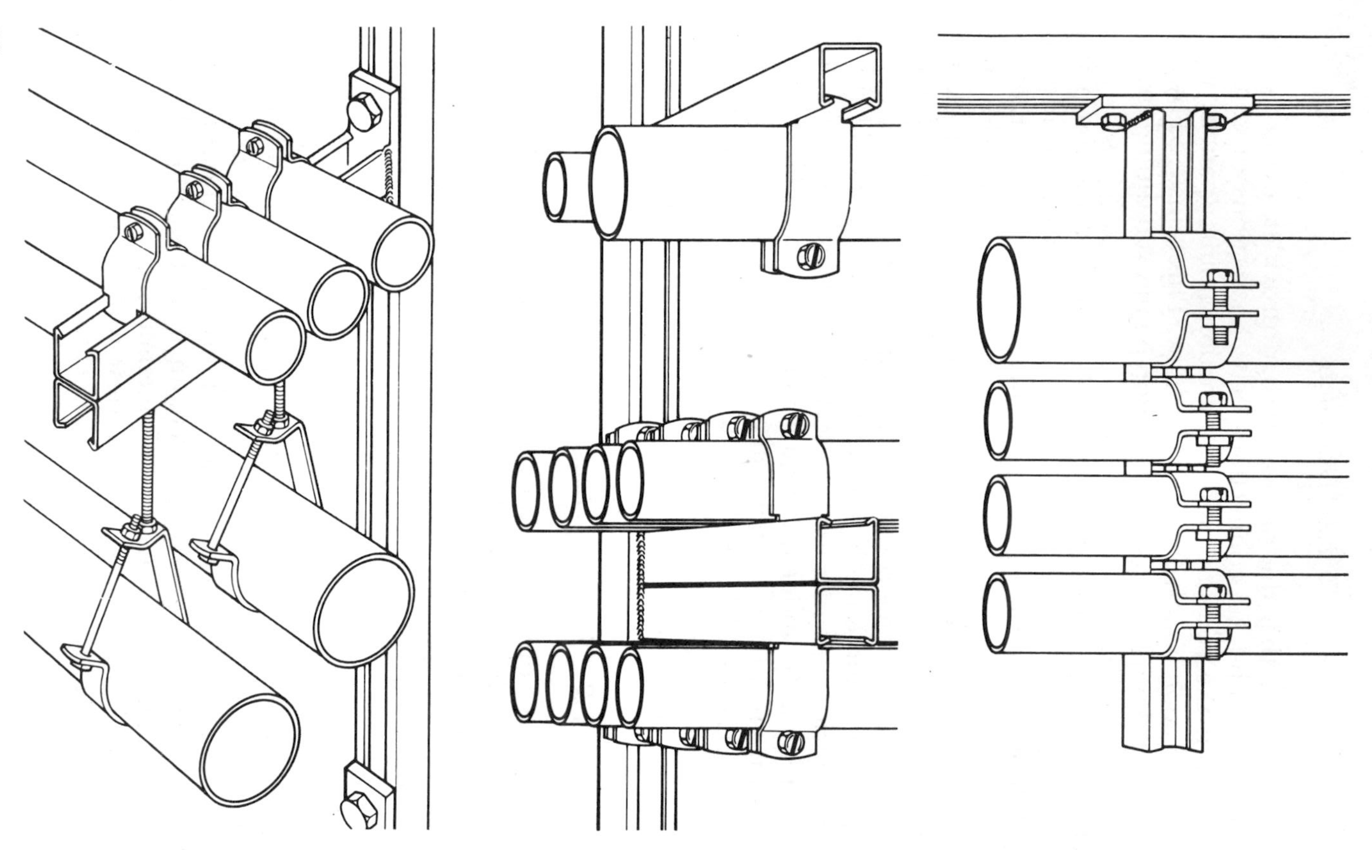

Figure 7.4. Several applications of metal support channels and single-bolt pipe straps.

Raceways suspended singly below ceilings or other structural members such as beams, columns, or purlins may be supported by lay-in pipe hangers as shown in Figure 7.5. Note that all of these examples require no drilling, welding, or fastening by means of power actuated tools. Such hangers can save the contractor much labor in installing surface-mounted raceways. For example, the column mount support in Figure 7.6 will lower installation costs by providing a quickly installed method of mounting pipe, conduit, switch, and outlet boxes or panel boards within the web of a column. Hardened steel set-screws bite into the column and hold securely, eliminating costly welding or drilling. In addition, a built-in high safety factor has reduced engineering time and utility of design has eliminated on-the-job improvising.

On new construction, channel inserts (Figure 7.7) are often embedded in concrete and such practice provides for many installation economies, both in initial construction and when up-dating of the mechanical and electrical system is required. The initial economies of construction stem from the ease with which pipe, air conditioning, lighting, and other fixtures can be attached to ceilings or walls.

Inserted by casting into the structure, continuous-slot channels accept all the assembly parts and fittings of the system. This provides virtually limitless structural arrangements—present and future. The installation of continuous-slot channels offers an immediate savings in time and labor by eliminating the need for precise calculation and measurement, both in layout planning and actual installation of attachment devices. There are additional savings in time and labor because changes or additions can be made readily to the existing channel at any time. The need for costly drilling in concrete and other costly procedures can be eliminated.

It is usually not practical to make an actual count of fastenings that are required for the installation of hangers, supports, and so on, as this would be too time-consuming. However, with a knowledge of the footage of various sizes and types of raceway under the different installation situations, the contractor or his workers should be able to make an educated guess of the type and quantities of the required fastenings.

Although an exact count of fastenings is not advised, the matter of obtaining the correct quantities should not be taken lightly. A considerable amount of lost time can accrue when the necessary fastenings are not available when they are needed.

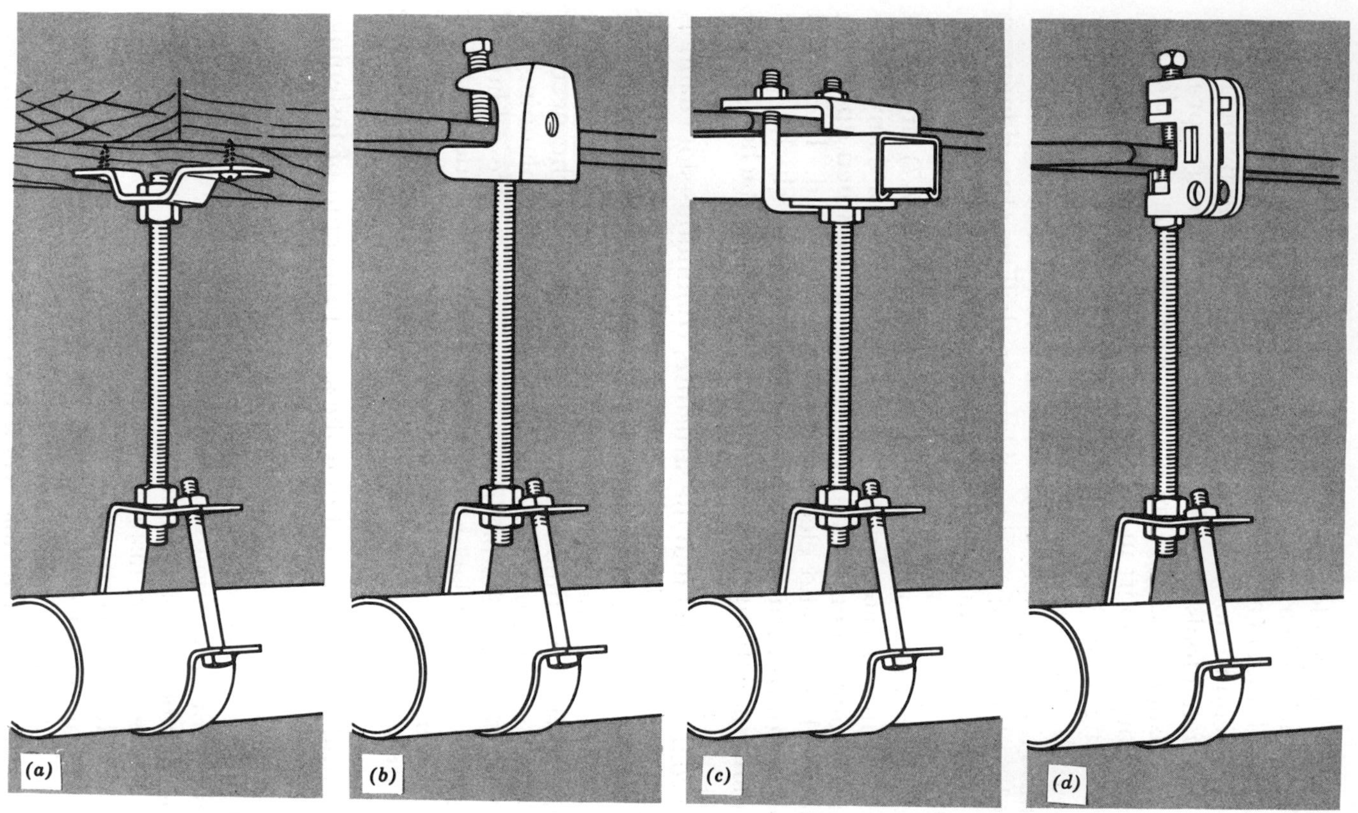

Figure 7.5. Raceways suspended singly below ceilings or other structural members such as beams, columns, or purlins may be supported by lay-in pipe hangers.

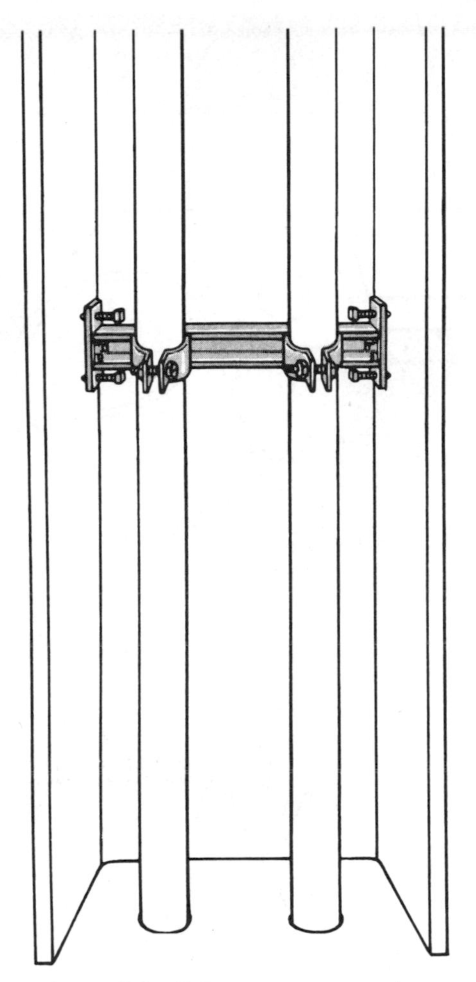

Figure 7.6. Column mount support.

7.3 Service and Feeder Wiring

Large sizes of conductors and cables are usually shipped on reels, which involves considerable weight and bulk. Setting up for a cable pull (transporting reels of cable to the desired location, measuring and cutting required lengths, etc.) will often account for a relatively large amount of the total cable installation labor consumed on this phase of the system. It is therefore necessary

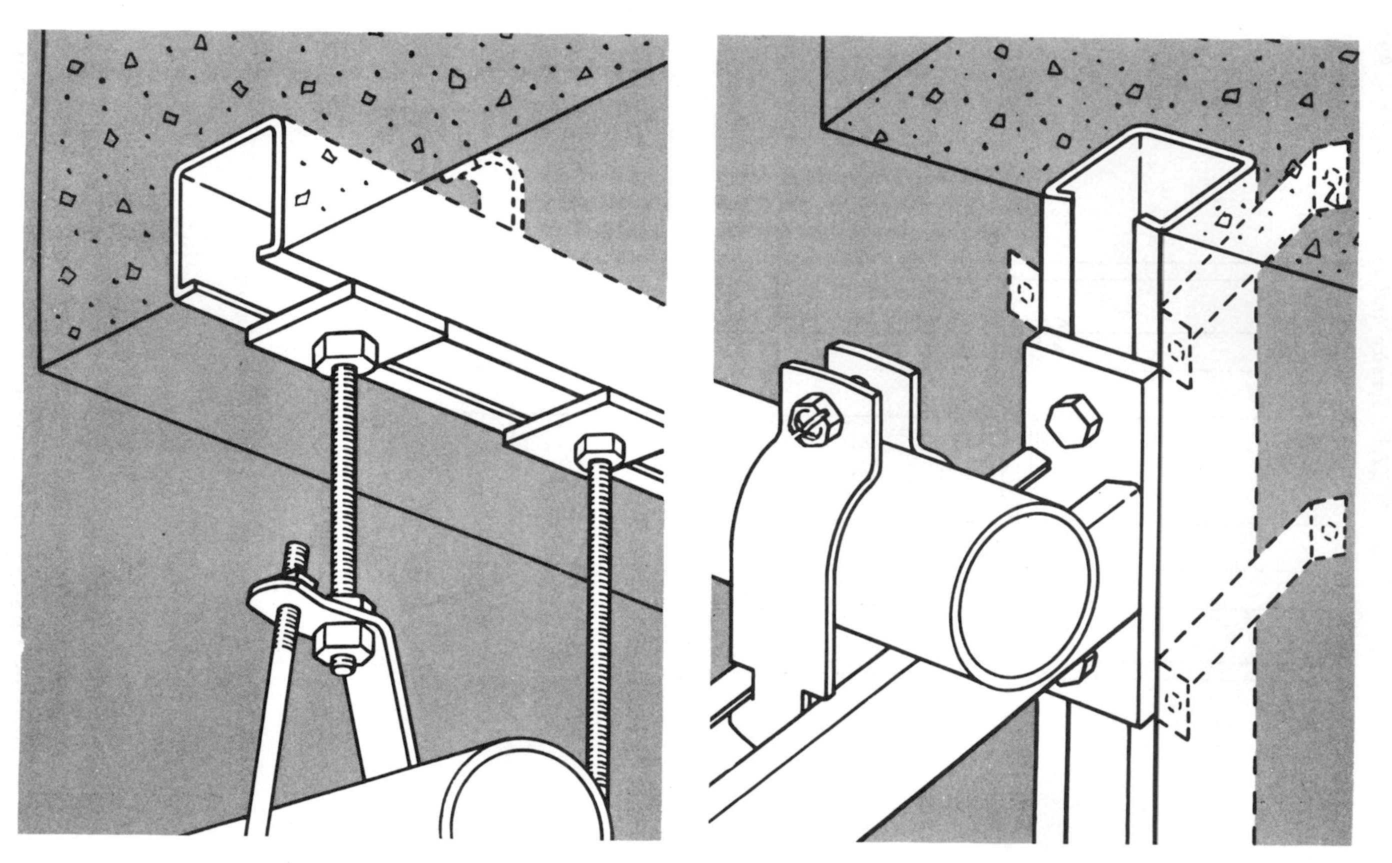

Figure 7.7. On new construction, channel inserts are often embedded in concrete and such practice provides for many installation economies.

that the proper equipment and organization be used so that the minimum amount of time will be consumed.

Whenever possible, the conductors should be pulled directly from the reels without prehandling them. This can be accomplished by ordering conductors pre-cut to the required length and then wound on the reels—at the factory—in three or four conductor pulls. Of course, this requires extremely close checking of the drawings and adequate allowances for lengths of conductors in the raceway system. The charge for this service by the factory will also be slightly higher than if the conductors were purchased in one-conductor reels. However, the savings in labor cost on the job will normally more than offset the cost of such coordinating effort.

As an extra precaution against errors in calculating the lengths of conductors involved, it is good practice to actually measure all runs with a fish tape before starting the cable pull, checking the totals against the totals indicated on the reels. When the feeder raceways have been installed at a relatively early stage of the overall building construction, it may not delay the final completion of the electrical installation to delay ordering the cables until the raceways can actually be measured.

Each job will have to be separately judged regarding the best location of pulling setups in order that the number of setups be reduced to a minimum in line with the best direction of pull. For example, it is usually best to pull cables downward rather than up, to avoid having to pull the total weight of the cable at the final stages of the pull and to avoid the possibility of injury to workers should the conductors break loose from the pulling cable on long vertical pulls. However, if heavy feeder cables are to be pulled downward for any appreciable length, the cable reel must be fitted with some type of brake so that the reel will not turn too fast because of the dead weight of the vertical cable runs. Also, the inner ends of the cables (at the reel core) must be anchored so that the ends do not get away from the pulling gang when approaching the end of the pull.

Conductors that are to be installed downward should be fed off the top of the reel and where conductors are to be fed upward, the best method is to feed from the bottom of the reel. This eliminates sharp kinks or bends in the conductors.

When the conduit is installed through a pull box, it should enter and leave the box in such a manner as to allow the greatest possible sweep for the conductors. Large conductors, especially, are difficult to bend and by proper planning, the electrician can simplify the feeding of these conductors from one conduit to another. Hydraulic and ratchet cable bending tools are now available which will quickly and easily bend cable from 350MCM to 1000MCM.

Consideration must also be given to conductor supports in vertical raceways. Article 300-19 of the NEC requires that one cable support be provided at the top of the vertical raceway or as close to the top as practical, plus a support for each additional interval of spacing as specified in the table, "Spacing for Conductor Supports." Such devices can save much time in handling large sizes of conductors as they are designed to make one-shot bends up to 90° and then automatically unload the cable.

SPACINGS FOR CONDUCTOR SUPPORTS

Wire Size	Conductors	
	Aluminum	Copper
No. 18 to No. 8 . . . . . not greater than	100 feet	100 feet
No. 6 to No. 0 . . . . . not greater than	200 feet	100 feet
No. 00 to No. 0000 . . . . . not greater than	180 feet	80 feet
211,601 CM to 350MCM	135 feet	60 feet
350MCM to 500MCM	120 feet	50 feet
500MCM to 750MCM	95 feet	40 feet
Above 750 MCM	85 feet	35 feet

Compact portable power pullers are available that allow only one man to pull a considerable amount of cable in a raceway system. Most of these tools weigh less than 50 pounds and are designed for set up in less than 5 minutes for use on 1¼ through 2½-inch conduit. A front switch on the puller leaves both hands free during the pull.

Other heavy-duty cable pullers are available for larger pulls that can save from 30–50% in labor over the other pulling method.

When pulling cable through manholes or on cable tray system, hook-type cable sheaves should be used (Figure 7-8) to change the direction of the cable under pull or to suspend the cable on long spans or to keep the cables bunched together for overhead pulls.

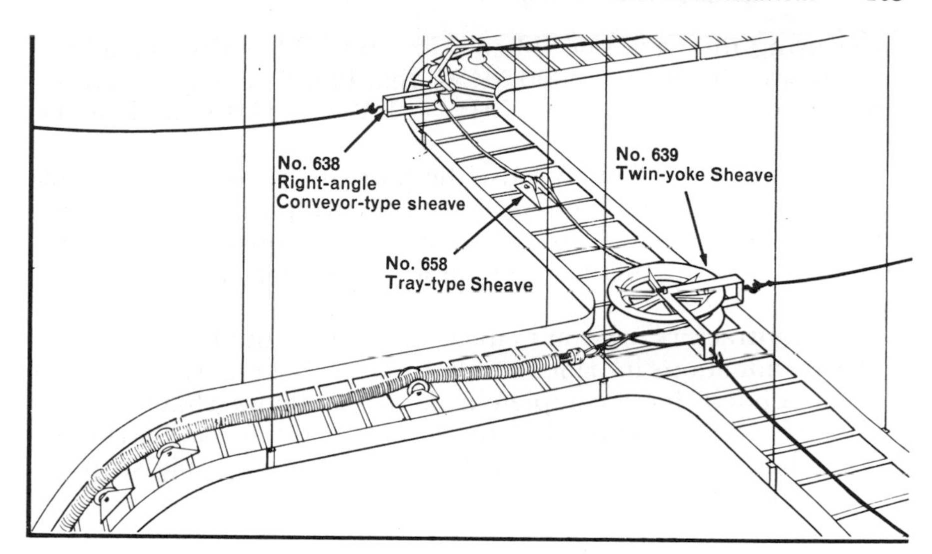

Figure 7.8. When pulling cable through manholes or on cable tray systems, hook-type cable sheaves should be used to change the direction of the cable under pull or to suspend the cable on long spans or to keep the cables bunched together for overhead pulls.

7.4 *Terminations*

The termination of the conductors at points of connection must be done with care in order to prevent expenditure of unnecessary time repairing the results of poor connections. In making such connections, three factors are involved: (1) tightness of the connection, (2) area of contact, and (3) insulation at the point of termination.

Looseness of connections or limited area of contact will cause heating at the point of connection which will not only reduce the efficiency of the circuit, because of the resulting higher resistance, but will actually damage the conductor insulation and the equipment. Proper size and type of connecting lugs, securely tightened, must be used.

Again, there are several aids available for the electrical contractor which facilitate the termination of large sizes of service and feeder conductors. First of all, cable cutters, capable of cutting conductors through 1000 MCM, save the workmen much time over using the hack saw. Then there are cable strippers, adjustable from 1/0 through 1000 MCM, that handle midspan and termination stripping of TW, THW, THWN, THHN, and similar insulation.

To use the tool, simply close the jaws of the tool on the cable and twist. These self-feeding devices assure positive progression down the cable to any position desired. To stop stripping, apply back pressure to the stripper until a full circle has been completed.

For contractors who anticipate a great amount of large-size cable terminations in panelboards, junction boxes, and so on, it is recommended that a complete make-up kit be purchased. These kits have everything needed for slicing, bending, and stripping cable; that is, cable cutter, ratchet cable bender, cable stripper, and a carrying case.

8

Special Systems

The job management techniques covered in this chapter deal with such special systems as electrical wiring in hazardous locations, main distribution panels, motor control centers, underfloor ducts, electric heat, underground systems, and the like. The use of efficient practices in the installation of these systems can greatly reduce the amount of labor consumed.

In laying out special electrical systems for the workmen, the electrical contractor or his supervisory personnel must make sure that all necessary items of material are available and that the workers are familiar with installation procedures for the system in question. More often than not, it will be found that all or part of the project's electricians may have little or no previous experience with the installation of certain special electrical systems. When this condition exists, time should be taken to explain the working procedures prior to beginning the installation.

8.1 Hazardous Location Wiring

Any area in which hazardous concentrations of flammable gases, vapors, or combustible dust is present is considered as a hazardous location. In all such areas, explosionproof electrical equipment and fittings designed to provide an explosionproof wiring system must be used.

The wide assortment of explosionproof equipment, fittings, and lighting fixtures now available make it possible to provide adequate electrical installations under these hazardous conditions. The basic principle of explosionproof wiring is to design the equipment and installation so that when the inevitable sparking or explosion occurs, the ignition of the explosive atmospheres surrounding the wiring system by sparks, flashes, or explosions within the equipment or system is prevented.

Hazardous locations have been classified in the NEC into certain class locations and various atmospheric groups have been established on the basis of the explosive character of the atmosphere for testing and approval of equipment for use in the various groups.

CLASS I LOCATIONS. Class I locations are those in which flammable gases or vapors are or may be present in the air in quantities sufficient to produce explosive or ignitible mixtures. Examples of this type of location would include interiors of paint spray booths where volatile flammable solvents are used; inadequately ventilated pump rooms where flammable gas is pumped, drying rooms for the evaporation of flammable solvents, etc.

CLASS II LOCATIONS. Class II locations are those which are hazardous because of the presence of combustible dust. Examples of Class II, Division 1 areas include working areas of grain handling and storage plants, rooms containing grinders or pulverizers, and all other areas where combustible dust may, under normal operating conditions, be present in the air in quantities sufficient to produce explosive or ignitible mixtures. Class II, Division 2 locations are areas where dangerous concentrations of suspended dust would not be likely, but where dust accumulations *might* form.

CLASS III LOCATIONS. These locations are hazardous because of the presence of easily ignitible fibers or flyings; but such fibers or flyings are not likely to be in suspension in air in quantities sufficient to produce ignitible mixtures. Such locations usually include some parts of rayon, cotton, and other textile mills, clothing manufacturing plants, woodworking plants, and establishments involving similar hazardous processes or conditions.

In general, rigid metallic conduit is required for all hazardous locations except for special flexible terminations and as otherwise permitted in the NEC. The conduit should be threaded with a standard conduit cutting die which provides ¾-inch taper per foot. The conduit should then be made up wrench tight to minimize sparking when fault current flows through the conduit system. Where it is impractical to make a threaded joint tight, a bonding jumper should be utilized. All boxes, fittings, and joints should be threaded for connection to the conduit system, and should be of an approved explosionproof type. Threaded joints should be made up with at least five threads fully engaged. Where it becomes necessary to employ flexible connectors at motor or fixture terminals, flexible fittings approved for the Class location must be used.

Seal-off fittings are required in conduit systems to prevent the passage of gases, vapors, or flames from one portion of the electrical installation to another through the conduit. Sealing compound used with the seal-off fittings must be approved for the purpose and should not be affected by the surrounding atmosphere or liquids, and should not have a melting point of less than 200°F.

Most sealing-compound kits contain a powder in a polyethylene bag within an outer container. To mix, remove the bag of powder, fill the outside container with water up to the marked line on the container and pour in the powder and mix.

Most other explosionproof fittings are provided with threaded hubs for securing the conduit as described previously. Typical fittings include switch and junction boxes, conduit bodies, uniforms and connectors, flexible couplings, explosionproof lighting fixtures, receptacles, and panelboard and motor starter enclosures.

8.2 Panelboards

Historically, the delivery of large panelboards, switches, and similar service equipment has been rather slow. Therefore, the electrical contractor and his supervisory personnel should make certain that this equipment is ordered shortly after the contract has been awarded, and then follow-up on the order to make sure the equipment is delivered on time.

The electrical contractor must make certain that especially the panelboard cabinets ("cans" as they are often called) are delivered to the job site in plenty of time. In most cases, these cabinets must be installed during the time the walls are erected and all conduit must be run and attached to these cabinets prior to closing in the building. If there is an anticipated delay on the complete panelboards, the electrical contractor should obtain a guarantee that the cans will be shipped at a specified date. Then the interiors may be installed later.

When ordering service equipment, their physical dimensions with relation to doorways, beams, ceilings, and space available must be given careful consideration. The dimensions may be checked by reference to catalog information in the case of standard equipment, or to shop drawings prepared by the manufacturer in the case of custom-built equipment.

Weight in relation to handling and moving equipment over floors, floor covering, and installation at a point above the floor level must also be given consideration. While the physical dimensions of such equipment may be such that it could be moved into place as a completely assembled unit, the weight of the complete assembly is often such that it would have to be handled in sections.

In some instances, it may be necessary to move such equipment over floors that are not of sufficient strength to bear the weight load of the complete assembly, or over floor coverings which would be damaged in moving the equipment over it. In such instances, the equipment would have to be handled in sections or adequate provision made for additional support or protection of the floor surface.

In some instances, particularly in industrial plants, it may be required to install such equipment at some height above the floor level. The weight of the equipment, as well as its physical dimensions must be given consideration in arranging for installation facilities and in determining the adequacy of permanent supporting structure of greater strength provided than was originally designed in order to preclude the possibility of failure of the supporting structure and resultant damage to the equipment and injury or death to persons.

The electrical contractor should be especially concerned about the preceding condition if any changes have been made in the original working drawings. For example, during a plant addition where heavy electrical equipment is required to be installed at some distance from the floor and be supported on the building's structural members, the plant maintenance supervisor may request that the equipment be relocated to a different area for easier access. This slight move may not appear to cause any problems, but always check with the architect or engineer in charge of the project to determine if the supports are adequate. If the plant's own engineering department has designed the system

Figure 8.1. In some instances, particularly in industrial plants, it may be required to install switches and other such equipment at some height above the floor.

and there is any doubt in the electrical contractor's mind about the possibility of failure of the supporting structure, it would pay the contractor to commission a registered structural engineer to examine the situation.

When panelboards are received, packing boxes should be immediately examined for damage and if such exists, this condition should be pointed out to the truck driver who can report the fact to the trucking company. Boxes should be stored in a dry and clean location until panels are installed. When it is not possible to install fronts of panels at the same time as the interiors, the fronts should be left in the packing boxes until they are installed. The installation of panelboards, service equipment and load centers is a specialty job. Experienced mechanics should be assigned to this type of work.

When boxes are located on the surface of existing walls, the walls should be either flat or the low points shimmed out so that boxes are not strained out of shape when secured to walls. For flush boxes, the boxes must be securely fastened and made both plumb and flush with finished wall surfaces.

No panelboard interiors should be installed until all of the wires have been pulled in and the plastering finished around the box. Before panelboard interiors are installed, the plaster and building grit should be cleaned off of the wires, the front fasteners and out of the box.

Panelboards should be carefully centered and properly plumbed. Panelboards must also be adjusted outwardly so that the opening in the front makes a close fit with the edges of the panelboard and at the same time the front must be flat against the finished plaster or wall. As a final step, directory cards on the inside of panel covers should be marked with the locations of all branch circuits.

While the use of metal enclosed panelboards is increasing, assemblies of individual enclosed externally operated switches or circuit breakers, interconnected by nipples or gutter connectors, are used.

Such assemblies are normally mounted on a backboard of wood sheathing. Some will require the backboard to be painted or covered with sheet metal prior to mounting the equipment units

Figure 8.2. While the use of metal enclosed panelboards is increasing, assemblies of individual enclosed externally operated switches, interconnected by nipples or gutter connectors, are used.

on it. In special cases, the specifications may require that the assembly be mounted on steel plate or on an angle or channel iron structure.

In most cases, the general arrangement of the equipment and their space requirements should be given consideration before the installation begins. Most contractors prefer to make a scale drawing of the layout to get the best arrangement and to make certain that the components will fit into the space provided. Several schemes should be tried in order to determine the best arrangement with respect to neatness and efficiency.

Once the general arrangement has been laid out, the backboard should be constructed out of channel, angle iron, plywood, or a combination of these materials. The equipment is then secured to the backboard as per the sketch discussed previously and interconnecting nipples, connectors, wire trough, and so on are installed as required.

The feeder conductors may then be pulled into the trough, nipples, and to the disconnect switches so that terminations may be completed. At this time, each feeder should be identified and labeled. Install proper size fuses, check for ground faults, and the system is ready to be energized, at least to the load side of the switches.

Certain types of equipment may often be more applicable to a given situation than others, so the electrical contractor and/or his supervisory personnel should give some thought to the type, as well as the manufacturer, of service equipment before such items are ordered.

8.3 Motor Control Centers

Motor control centers are normally specially built or assembled at the manufacturer's plant and then shipped to the job site in one or more sections depending upon the size and weight of the final assembly.

When the assembly reaches the job site, organization with the electrical workers as well as the general contractor is necessary in order to transport such equipment from the point of delivery into its final location within the building. On most projects cranes will be available, and when a crane, forklift, or similar apparatus can be rented, this is the best way to handle the moving of such heavy equipment. Should a crane not be available at the job, the electri-

Figure 8.3. Motor control centers are normally specially built or assembled at the manufacturer's plant and then shipped to the job site in one or more sections, depending upon the size and weight of the final assembly.

cians will have to resort to the use of rollers, chain hoists, hydraulic jacks, and perhaps a cable pulling machine to move the equipment into its final location.

Motor control centers normally require a great number of feeders and other conduits as well as bus duct, cable trays, and similar systems entering and leaving the center. This can present serious space problems above and below the unit. Therefore, proper advance planning and coordination of the raceway runs in relation to the motor control center will result in time saved in making the installation.

The preparation of working and shop drawings showing the layout, dimensions, and so forth of such equipment will also reduce the on-the-job time consumed in laying out and installing the equipment and raceways.

8.4 Underfloor Ducts

Underfloor ducts, either steel, fiber, or PVC plastic are often used in electrical systems for building construction, especially where a grid system for future service is desired.

Underfloor raceways are almost always installed concealed in concrete floor slabs and since the components are normally not "stock" items, the electrical contractor must make certain that all necessary materials are on hand prior to beginning the installation. The items are seldom available at local suppliers and any delay in the concrete pour caused by shortages of materials will most certainly cause the electrical contractor much grief—either in having to pay for damages or else through a damaged reputation.

Many electrical contractors may carefully calculate the required underfloor ducts, connectors, and the like, but will overlook some of the accessories that will be required for a complete installation of the system. These seemingly incidental accessories are just as important as the main ducts and can also cause a delay in the concrete pour. For example, special brackets may be needed on a particular project using an underfloor system to support raceway or related boxes at the proper height from the deck. These items must be anticipated and arrangements made for delivery prior to starting the installation. A close check of the duct manufacturer's catalog will help the contractor determine exactly what will be required for any particular project.

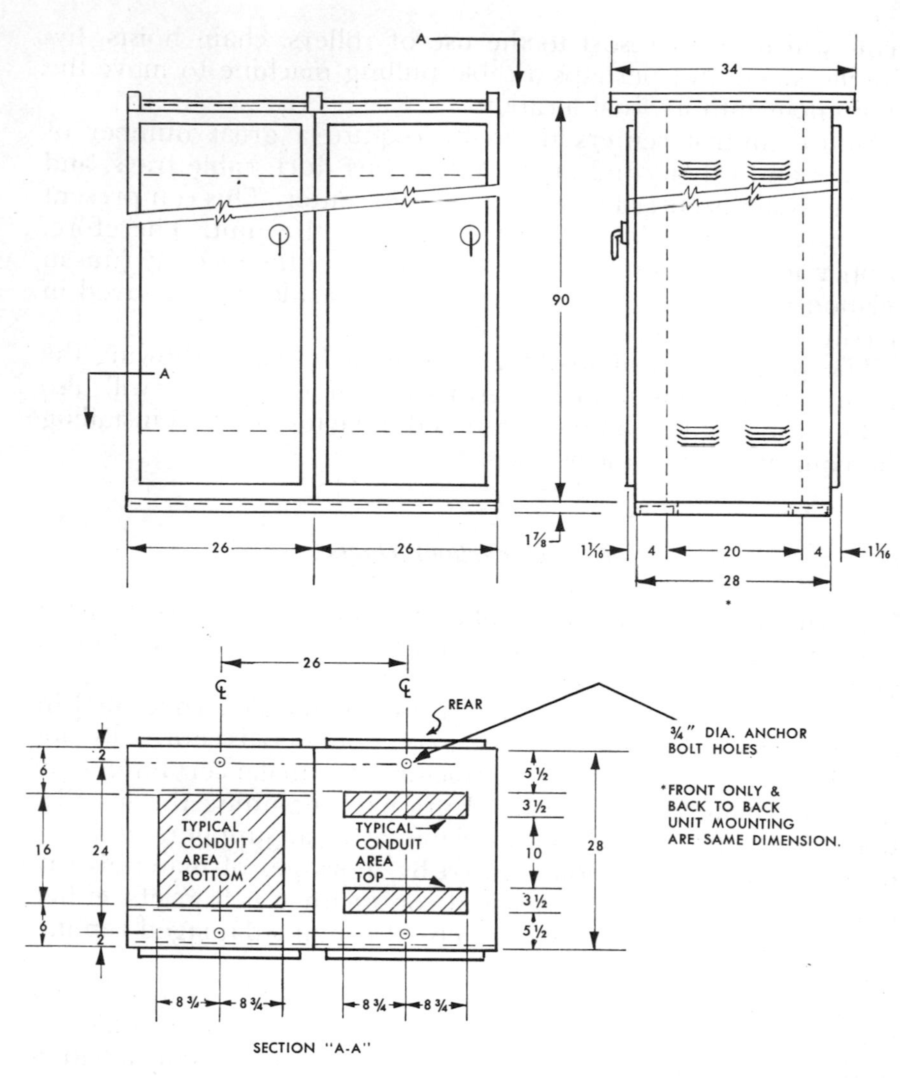

Figure 8.4. The preparation of working and shop drawings showing the layout, dimensions, and so on of motor control centers will greatly reduce on-the-job time consumed in their installation.

Since there is a critical relationship of the position of the underfloor ducts and their related junction and outlet boxes, the electrical contractor should seek assistance from the general contractor in verifying the exact elevation of the finished floor and in using a surveyor's level while the electricians are installing the boxes at the proper height.

Those in charge of the electrical work at the job site must also ascertain that the underfloor raceway is so arranged that there will be no difficulty encountered in pulling in the conductors at a later date.

In some types of underfloor systems, metal ducts are installed in the floor during the concrete pour and outlet box holes are later drilled with a diamond-head rotary drill bit. In such cases, the electrical contractor and his workmen must make notes of exact locations of these ducts for future reference.

8.5 Electric Heating

Over the past couple of decades, electric heating has found its way into thousands of homes throughout the United States, and in most cases, the electrical contractor is responsible for furnishing and installing the equipment.

Besides space heating equipment consisting of electric baseboard heaters, forced-air heaters, duct heaters, heating cable, and the like, the electrical contractor is also responsible for furnishing and installing other electric heating devices such as snow melting cable and mats, roof and gutter heating cable, and perhaps room heating and cooling units.

In order for such systems to be efficient and economical, and provide the desired comfort condition, they must be designed on the basis of precise heat-loss calculations of the areas to be heated. In most cases, these calculations are performed by consulting engineers who in turn provide complete working drawings and specifications detailing exactly how the system is to be installed.

In some instances, local power companies will help an electrical contractor calculate the heat loss for a given residence and then help to lay out the system, select the equipment, and so on.

Good working drawings and specifications should include the electrical characteristics of the installation, instructions for the selected heating equipment, and also a concise description of the selected heating and cooling controls, and should indicate the location of each. Since the location of the heating units within a particular area is all important in obtaining a comfort condition within the area, their positioning should be verified with the architect/engineer prior to the actual installation.

Equipment schedules may be used to identify the equipment from symbols and other identifying marks on the working drawings. The schedule should include the manufacturer, catalog number, approximate dimension of the units, the equipment rating (in watts), and other pertinent data. The electrical contractor and his workers should make certain that all of this information is completely understood before and during the installation.

At times, especially on residential projects, the electrical contractor may be called upon to calculate the heat loss of a building or area within a building, select the heating equipment, and make the installation. When such a situation arises, the goals of the electric heat installation are to obtain:

1. Adequate, dependable, and troublefree installation.
2. Year-round comfort.
3. Reasonable annual operating cost.
4. Reasonable installation cost.
5. Systems that are easy to service and maintain.

The person designing the system must make heat-loss calculations to ascertain that heating equipment of proper capacity has been selected. Heat loss is expressed either in Btu per hour (which is abbreviated Btuh) or in watts. Both are measures of the rate at which heat is transferred and are easily converted from one to the other:

$$\text{watts} = \frac{\text{Btuh}}{3.4}$$

$$\text{Btuh} = \text{watts} \times 3.4$$

Basically, the calculation of heat loss through walls, roof, ceilings, windows, and floors requires three simple steps:

1. Determine the net area in square feet.
2. Find the proper heat-loss factor from available tables.
3. Multiply the area by the factor; the product is expressed in Btuh. Since most electric heat equipment is rated in watts, divide this product by 3.4 to convert to watts.

Calculations of heat loss for any building or area may be made more quickly and more efficiently by using a prepared form with spaces provided for all necessary data and calculations; the procedure then becomes routine and simple. Such forms and a complete explanation of their use are available from manufacturers of heating equipment. Check with your local sales representative or supplier for details on how to obtain these forms.

Load estimate is based on desing conditions inside the building and outside in the atmosphere surrounding the building. Outside design conditions are the maximum extremes of temperature occurring in a specific locality. The inside design condition is the degree of temperature and humidity that will give optimum comfort.

8.5.1 Controls for Electric Heating Equipment

When ceiling heating cable or electric baseboard heating equipment is utilized, each room or separate area should be temperature controlled by its own wall-mounted thermostat. The thermostat should be mounted 48–54 inches above the finished floor and should always be kept away from heat-producing equipment. All 240-volt heaters should be the two-pole type with an OFF position, or low-voltage thermostats may be used in conjunction with relays.

Fan-driven wall heaters can be controlled by either wall-mounted or build-in thermostats. While electric baseboard heaters may also be controlled with thermostats built into the unit, this practice is not recommended since control is seldom accurate.

When electric heat is supplied by a duct system in the form of electric heating elements in a forced-air furnace or ductwork, the work is normally handled by a heating and cooling (mechanical) contractor. However, the electrical connections are almost always handled by the electrical contractor.

Circuit requirements for this type of heating equipment do not greatly differ from those for any other circuit, but the electrical contractor and his electricians must carefully study the manufacturer's electrical wiring diagrams for the equipment in question to determine exactly the number and type of circuits required for the system.

8.6 Underground Systems

Underground installations consist of cables buried directly in the earth or else pulled through an underground raceway system.

The cables themselves are used for transmission and distribution facilities and also for signal and communication systems.

Direct burial installations will range from the installation of a relatively small single-conductor cable to large multiconductor distribution cables. However, in both cases, the conductors are installed in the ground either by placing them in an excavated trench which is then backfilled, or by burying them directly by means of a cable plow which opens a furrow, feeds the conductors into the furrow and closes the furrow over the conductor.

Where the job site conditions permit, the use of the cable plow is by far the fastest way to install direct-burial cable. There are certain condition, however, that will prevent the use of such a plow; they are as follows:

1. The type of earth in which the cables are to be buried.
2. The types of conductors.
3. Presence of existing underground facilities.
4. Nature of the terrain.
5. Contract specifications.
6. Amount of work involved and the availability of the equipment.

When the conductors are to be installed in a trench, the requirements of the contract would be carefully noted for special installation requirements such as the placing of a layer of sand in the bottom of the trench, removal of stones or rock from the backfill, or the placing of creosoted wooden boards over the conductors for protection during backfilling.

Underground raceway systems consist of manholes or junction boxes and connecting runs of one or more conduits (rigid, PVC, fiber, etc.) placed in trenches and sometimes encased in concrete. The size and number of raceways depends upon the use of the system, number of conductors, and the spare raceway capacity desired, but will normally range in conduit size from 2 to 6 inches, inclusive.

When an underground raceway system does not require large chambered manholes to serve the purpose of installing and splicing conductors and small items of equipment, relatively shallow junction boxes are often used. These are usually constructed of cast iron, steel or concrete with a steel square or oblong steel plate cover.

During the installation, underground raceways are usually installed so that the top tier of conduits are at least 2 feet below the finished grade. From this, it is easy to determine the overall depth of the trench by calculating the size of the conduits, the separation between layers, and the distance of the bottom conduit from the bottom of the trench.

The digging of the trenches for either an underground raceway or direct burial cable installation should be given careful consideration and planning. The electrical contractor should ascertain all of the potential ground and other conditions that will probably exist. The follow are some of the factors that should be considered:

1. Type and condition of the soil.
2. Presence of shale, stones and boulders, or rock formations that would require extensive drilling or blasting.
3. Presence of stumps, tree roots, and the like.
4. Presence of existing underground utilities.
5. Depth of the trench.
6. Type of grade surface: paved or unpaved.
7. Pedestrian and cars or other traffic conditions.

In the past, copper wire has been used almost exclusively for underground wiring systems because of copper's low resistance. Aluminum conductors have the advantage of lightness, but since it has a higher resistance than copper, aluminum conductors would be larger than copper cable, requiring larger ducts or conduit. In direct burial installations, aluminum wire is now used extensively, but in conduit or duct systems, copper is still used mostly. This is because the ducts are usually more costly than the cable and so must carry as much power as possible.

The installation of potheads on the ends of lead-sheathed and other high-voltage cables serve to exclude moisture from the ends of the cable, protect the cable against mechanical injury, and provide terminals for the cable conductors. When dealing with potheads and the splicing of high-voltage cable, only qualified cable splicers should be used to perform the work.

Large cables are ordered in lengths to fit the duct or trench sections, plus an allowance for slack and splicing. Smaller cables are shipped on standard reels and are cut on the job after they have been installed in the ducts or laid in the trench. Cables may be pulled by hand, although the use of a cable-pulling machine or winch is the most desirable method. Sometimes a saving in labor can be realized by ordering the cables of three- or four-wire systems made up on one reel, so that all cables are pulled together.

To prevent the spread of damaging arcs from a cable fault to adjacent cables in conduits, manholes, and handholes, the cables are wrapped with an asbestos tape saturated with sodium silicate, or each cable may have a cement covering applied in the manhole.

8.6.1 Special Tools for Underground Wiring

There are several tools available that facilitate the installation of underground wiring systems. All electrical contractors and their personnel should become familiar with these tools and learn how to use each to its best advantage.

Trenches, for example, are either hand-dug or machine dug. For short turns, hand digging usually suffices, but even the use of compressed air–operated jack hammers with appropriate drills, spade, and tamping accessories would greatly reduce the time and fatigue over using a hand pick and tamper.

Figure 8.5. A boring machine can save the contractor much time in installing raceways under sidewalks or drives.

In areas that do not have adverse ground conditions or the existance of other utilities crossing or in a line with the new trench run, the use of a trenching machine is more economical than hand digging. Self-propelled mechanical trenching equipment range from small hand-tractor type of trenchers to large riding trenchers. When the extent of the contractor's work does not warrant the purchase of a trencher, such equipment can usually be obtained on a rental basis.

For installing cable and conduit under roadways and sidewalks special boring machines are used to drill or push conduit under the walk or roadway so that neither is damaged. Such machines require only a narrow starting trench and terminal sump hole, making restoration of the area to its original condition less demanding. Drill heads range in size from 1¼ to 2 inches for the initial cutting pass, and from 2 to 3½ inches for the return or reaming pass.

When drilling under sidewalks or roadways is not practical, normally a suitable channel can be cut with a concrete saw. Two cuts are made with the saw along the path of the conduit or cable. This can be done with little or no damage to the surrounding pavement. The gravel and dirt is then removed to accept the conduit and cable, after which the cut needs only minor patching.

Other special tools include an electronic metal locator for locating underground cable and conduit; an underground fault locator for locating a break or ground fault in buried cable, and a power wire-pulling apparatus for pulling long runs of heavy cable through underground raceways.

9

Lighting Fixture Installations

Lighting designers strive to select lighting equipment that will provide the highest visual comfort and performance that is consistent with the type of area to be lighted and the budget provided. It is the electrical contractor's responsibility to see that all lighting equipment is furnished and installed exactly as selected and specified. To be able to do this, the workmen on the job should be furnished with a complete set of detailed lighting specifications along with comprehensive lighting plans and detailed drawings. If the drawings furnished by the architect or engineer are not detailed enough, the electrical contractor should provide additional drawings to supplement the project's working drawings.

This chapter covers installation procedures such as mounting lighting fixtures, relation of the lighting fixtures to the building type, special accessories, and other job management procedures with respect to lighting fixture installations. However, since lighting fixtures vary widely in quality, design and physical appearance, and so on, it would be impractical to try to cover all lighting systems for all projects. Instead, a selected sampling is given.

9.1 Factors Affecting Lighting Fixture Installation

The cost of lighting fixtures and their installation accounts for over 50% of the total contract cost on most commercial projects. For this reason, every effort must be made to organize and proceed with this phase of the electrical installation so that the expended labor can be kept to a minimum. To accomplish this, the following should be kept in mind by the electrical contractor, his superintendent, and his foremen.

Before starting any of the roughing-in for lighting circuits, those in charge of the project should become familiar with each type of lighting fixture through reference to the working drawings, written specifications, shop drawings, and so on. If the full information about each type is not available, proper inquiries should be made to the manufacturer to obtain the necessary information. This will enable the workmen to provide the correct outlets, hangers, frames, and the like; it will further insure that only the recommended number of fixtures will be connected to a circuit.

Make certain that the proper accessories are ordered at the same time as the lighting fixture. Many contractors have ordered only the housings for recessed fixtures, for example, and had planned to order the diffusers and trim at a later date to prevent them from becoming broken during the construction stage. However, when the order was finally made, they found out that the lighting fixture had been discontinued, causing them great expense trying to find the required number of trims from various suppliers around the country who happened to have a few left.

The electrical contractor should make sure that all lighting fixtures, accessories, components, lamps, and so on will be delivered in advance of their need. Not months in advance (unless delivery difficulties are anticipated), but within a reasonable time so as not to hold up the workmen or other trades working on the project. This may make it necessary to provide on-the-job storage facilities to house the lighting fixtures until they are needed; but in most cases, when the lighting fixtures are needed, there will be vacant space within the building where the fixtures can be stored until needed.

Very few projects will use lighting fixtures furnished by one manufacturer. Therefore, the fixtures will arrive on the job site at different times. When the shipments arrive at the job site, each should be checked by a responsible person to make sure that the proper type and quantity of each fixture has been received in satisfactory condition. When discrepancies are found, the proper persons should be notified at once to receive credit on damaged items and to reorder shortages.

Often on-the-job assembly and wiring of some of the lighting fixtures will be required. This is especially true of certain types of recessed fixtures and fluorescent fixtures. When such a situation exists, proper work space should be provided along with proper tools, accessories, and work benches. Items needed for the assembly, like fixture wire and lampholders, should be organized in the work area for greatest efficiency. Then the crew assembling the fixtures should receive proper instruction, and these same electricians should be left on the job until all fixtures are assembled.

At the time of installation, all necessary installation tools, ladders, rolling scaffolds, and so on should be delivered to the job. Avoid the need of tools and equipment halfway through a project; this only costs the contractor extra money.

Rolling scaffolds must be of the proper height to allow the men to work easily without having to stand on boxes or similar devices. They should be provided with bins at the working level to hold tools, fittings, wire nuts, connectors, tape, and other necessary accessories. The working platform itself should be large enough to accommodate at least two workers and a reasonable number of lighting fixtures. This prevents having to climb up and down the scaffold unnecessarily.

Well in advance of the fixture installation, the electrical contractor or his supervisory personnel should make arrangements with the general contractor and all of his subcontractors to have the area reasonably cleared of all equipment, trash, and so on to make the area accessible and to be able to roll the scaffolding without encountering unnecessary obstacles. Sometimes, however, a piece of fixed equipment will be in the area which cannot be moved. If the electrical contractor suspects that such a piece of equipment will hinder the lighting fixture installation, he should make all efforts to install the lighting fixtures prior to the installation of the equipment.

A well-organized rolling scaffold arrangement can contribute a great deal to the efficiency of making the installation. One electrical contractor in Washington, D.C., for example, rigged a rolling scaffold with a steering mechanism, chain and sprocket drive, and powered the apparatus with a ½ HP drill motor. This arrangement allowed the men working on the scaffold to power the scaffolding down a row of fixtures without needing an additional man on the ground to roll it.

When power tools are to be used in working from a scaffold, receptacles should be provided at the platform level. A single flexible cable extending from the scaffold to a power source should then be provided to avoid cluttering up the platform area with extension cords.

The job superintendent or foreman should keep a close check on the progress of fixture installations and immediately make any adjustments necessary to improve the work efficiency. This may require an increase or decrease in the work force, additional tools, additional hangers, equipment, or replacement materials as may be necessary.

If at all possible, the fixtures should be completely installed and wired in one area before proceeding to the next point of installation. This saves time in moving scaffolds, materials, tools, and the like from one area to another more than is necessary.

9.2 *Mounting Lighting Fixtures*

The height above the finished floor at which lighting fixtures are normally mounted could range from a couple of feet in the case of wall-mounted step lights to 20 or 30 feet in the case of high bay industrial structures. However, the average mounting height for most fixtures in commercial buildings will be around 9 or 10 feet. In the case of individual fixtures, a step ladder will suffice, but for rows of fluorescent fixtures, rolling scaffolds are still the best means of attaining the required height.

One of the main considerations for mounting lighting fixtures to obtain the highest installation efficiency is to select the proper hangers and accessories. Some careful thought in this area can save the electrical contractor sufficient man-hours to make the practice well worth while.

For example, concrete inserts (Figure 9.1) can be placed in concrete pours for use in hanging rows of fluorescent fixtures at a later date. The type shown in Figure 9.1 is manufactured with a knockout that saves covering or stuffing the opening during the concrete pours. Once the concrete has cured sufficiently, the knockout slots are removed and all-thread rod is inserted on which the fixtures are hung.

The T-bar clip in Figure 9.2 is another great time-saving device when installing surface-mounted fixtures to acoustical T-bar ceilings. This clip has an integral boss with a recessed 1/4-20 stud bolt. The clips are snapped on the T-bar channel quickly and then the fixtures are held in place by a cupped washer and wing-nut. Each clip will safely carry a load of 100 pounds. However, the electrical contractor must make certain that the T-bar ceiling is properly secured to hold the weight of lighting fixtures.

Other hangers include an adjustable swinging hanger flange which accepts ½-inch rod (Figure 9.3); I-beam clamps (Figure 9.4); and regular beam clamps (Figure 9.5). Channel bar is also frequently used to mount light fixtures as shown in Figure 9.6.

Figure 9.1. Concrete inserts can be placed in concrete pouts for use in hanging rows of lighting fixtures at some later date.

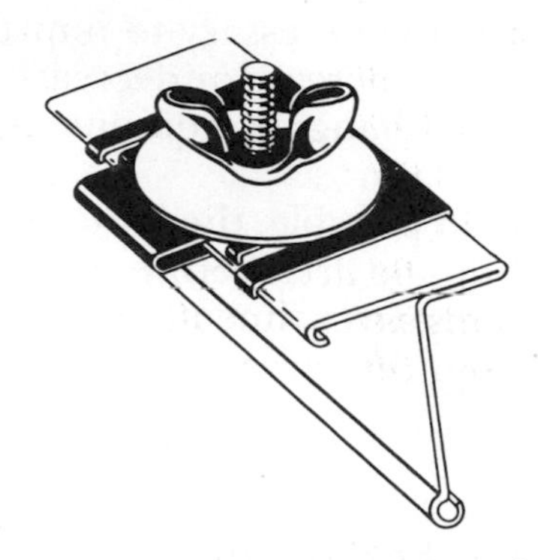

Figure 9.2. T-bar clips are another great time-saving device when installing surface-mounted fixtures to acoustical T-bar ceilings.

Figure 9.3. Adjustable swinging hanger flange.

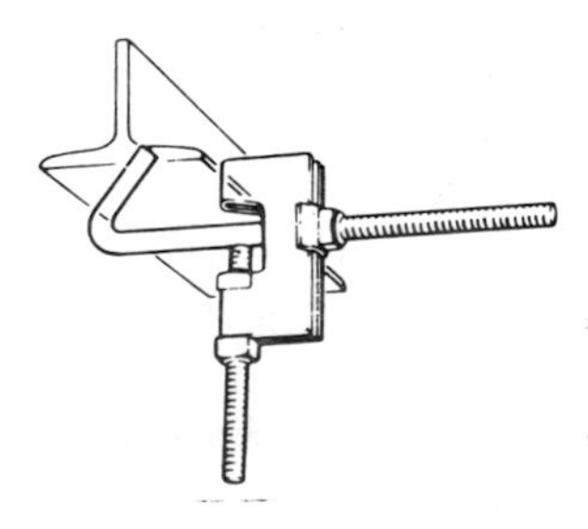

Figure 9.4. I-beam clamp.

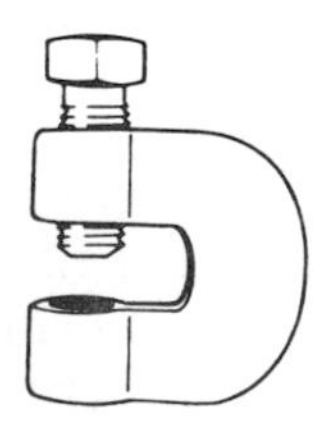

Figure 9.5. Regular beam clamp.

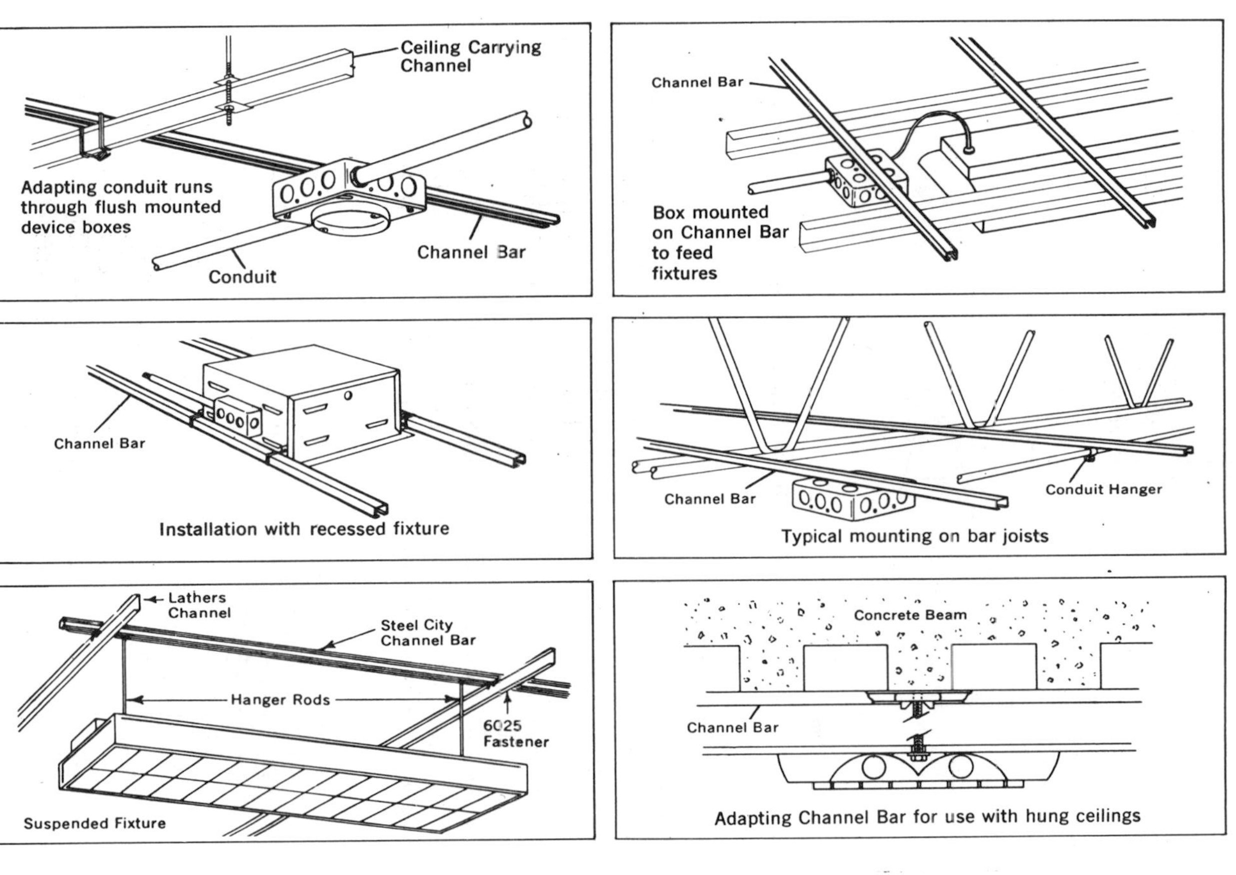

Figure 9.6. Channel bar is frequently used to mount lighting fixtures.

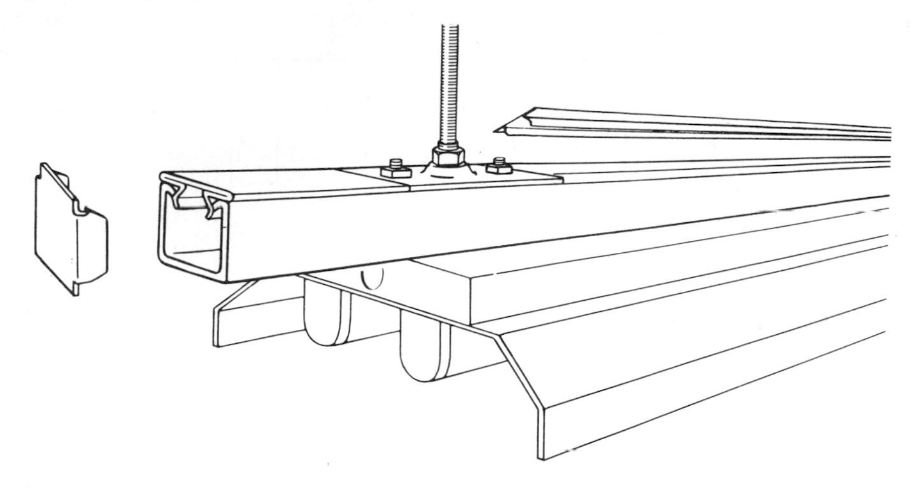

Figure 9.7. Channel system with fluorescent lighting fixture attached.

Channels for lighting systems can provide definite installation advantages on some projects. In most cases, this flexible system requires fewer attachments to the building structure and it incorporates certain built-in provisions for easy maintenance and future modifications when lighting fixtures must be added, deleted, or relocated.

A channel system offers the electrical contractor a means of simultaneously providing for the electrical feed and the mechanical support of lighting and other equipment. It further assures

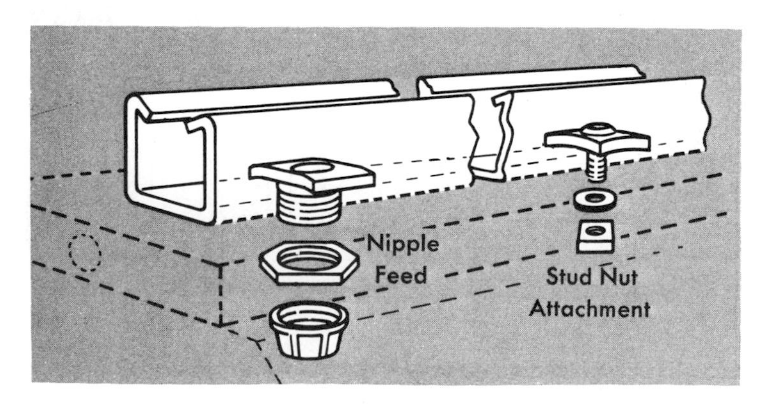

Figure 9.8. Methods of attaching lighting fixtures to channel.

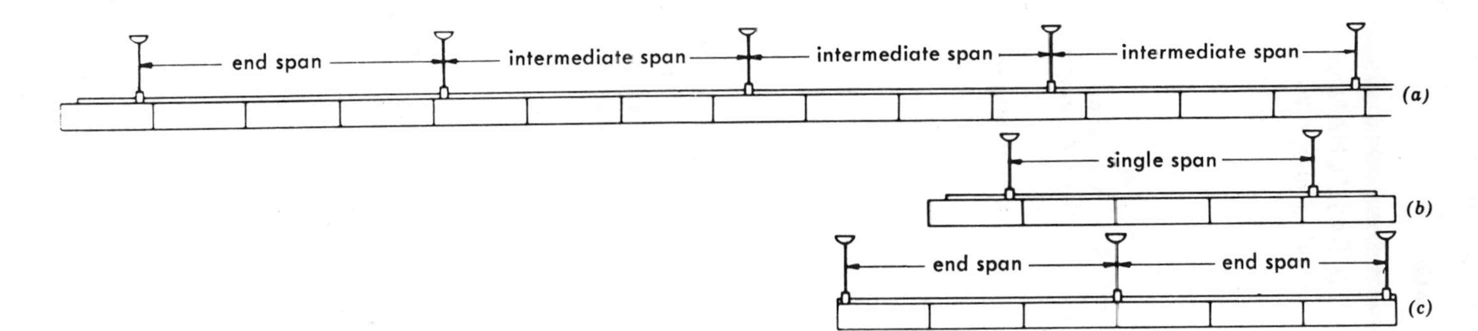

Figure 9.9. Example of different spans to calculate hanger distances. (a) Long continuous run; (b) single span; (c) double span.

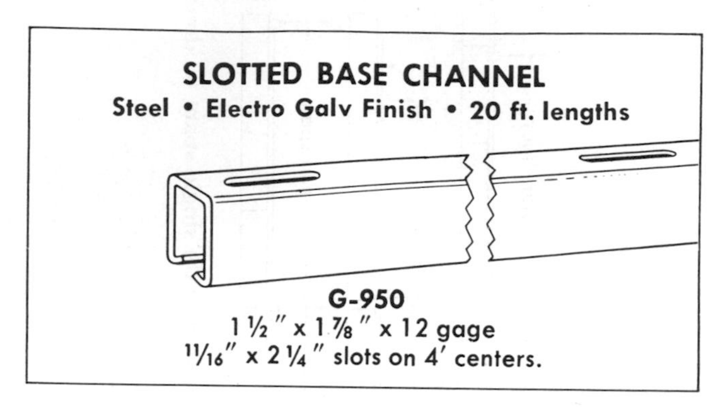

Figure 9.10. Example of slotted base channel.

true and rigid alignment and lends itself to systematic mass-assembly methods which economize on labor.

Figure 9.7 shows a channel system with a fluorescent lighting fixture attached while Figure 9.8 show the methods of attaching the fixture to the channel.

The hanger spacing is often determined by the type of building construction. The deflection then will determine the proper channel since this deflection should not exceed 1/240 of the span.

To estimate the deflection at the center of an intermediate span in long continuous runs (Figure 9.9), multiply the weight of a single fixture times the applicable deflection constant (Table in Figure 9.11). This deflection also applies to the end span in Figure 9.9a and the single span in Figure 9.9b if the dimension "C" is between ¼ and ⅓ of the length of the span. If a cantilever does

Deflection Constants for Continuous Run, 4-Foot Fixtures*.

span feet	B-906 B-956	B-900-M G-975-M	G-953	B-900 G-975	B-901 G-950, G-965	B-900-2A	B-902 G-955
6	.004	.000	.000	.000	.000	.000	.000
8	.009	.002	.001	.000	.000	.000	.000
10		.005	.004	.003	.001	.000	.000
12		.010	.007	.006	.004	.001	.001
14				.012	.007	.002	.002
16				.020	.011	.004	.004
18					.018	.007	.006
20						.010	.009

*For 8-foot fixtures reduce the deflection constant by 50%. This table is for normal weight fixtures—the constant ".000" infers negligible deflection.

Figure 9.11. Table of deflection constants for continuous rows of 4-foot fluorescent fixtures.

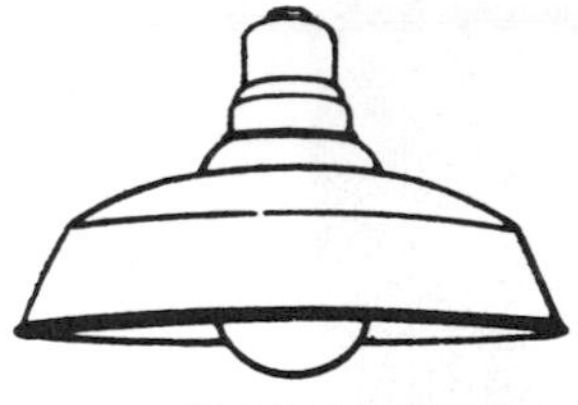

Figure 9.12. A dome-type reflector.

not exist as in the double span (Figure 9.9c), the deflection of end spans will be doubled.

9.3 Specialized Lighting

Lighting installations that may be termed "specialized lighting" include airport lighting, some types of exterior lighting equipment, and emergency lighting. The basic procedures that should be followed in providing efficient management of such installations include:

1. Be familiar with the installation and operating details of the system, its equipment, and components.

Figure 9.13. The electrical contractor must plan his lighting installation so that the fixtures do not interfere with heating, ventilating, and air-conditioning ducts and diffusers.

Figure 9.14. Recessed incandescent fixtures in permanent ceilings must be laid off carefully to avoid damaging the finished ceiling should a fixture have to be moved.

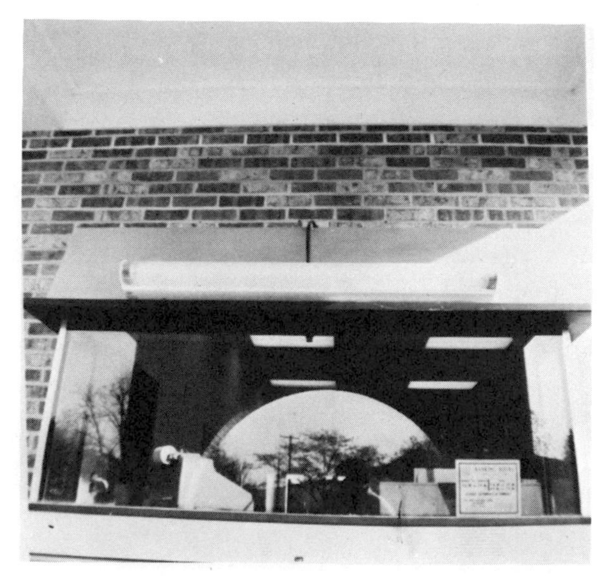

Figure 9.15. When installing lighting fixtures outside, the contractor should make certain that the fixtures are designed for outdoor use; that is, cold weather ballasts, raintight, and so on.

Figure 9.16. Examples of surface-mounted and recessed fluorescent lighting fixtures.

Figure 9.17. Outlet boxes roughed-in for pendant-mounted decorative lighting in modern church. Installations of this type take extra labor and must therefore be planned accordingly.

2. Have all items of material on hand prior to starting the installation.
3. Check all deliveries carefully and follow up on any discrepancies immediately.
4. Provide proper tools and installation equipment.
5. Provide for adequate and efficient fabrication and assembly facilities when needed.
6. Adequately inform the workmen of the installation procedures.
7. Maintain a careful check on the installation progress and try to complete each phase of the installation before starting another.

9.4 Prefabrication

When a considerable number of specific types of lighting fixtures are utilized on a project, each of which requires assembly, much time can normally be saved by assembling and wiring these fixtures at the shop or some central location on the job site. If several components are required for each fixture, like integral outlet or connection boxes, and stems, these should be prefabricated and preassembled at the same time.

A dome-type reflector, for example, as shown in Figure 9.12 may be specified to be hung 18 inches below the finished ceiling on swivel hangers. Such fixtures are normally shipped with the dome and lampholder unassembled. Much time can be saved by unscrewing the lampholder and wiring, cutting and inserting the stems, attaching the swivel base, and installing the dome reflector at one central location. Then when it is time to install the lighting fixtures, the preassembled units can be transported to the various locations and merely attached to the feeder wires with wire nuts and tightening the mounting screws on the outlet box. When lamped, the fixtures are ready for operation.

There are several other types of lighting fixtures that lend themselves to prefabrication. Those that fit into this category should become apparent to the electrical contractor upon reading the assembly and installation instructions of each fixture.

10

Personnel Management

Electrical contracting personnel normally fall into three categories: (1) management, (2) administrative, and (3) workmen. The management level consists of the top-level supervisory personnel such as the owner(s) general superintendents, estimators, and engineers. Administrative employees are comprised of bookkeepers, secretaries, and the like, while the workmen include job superintendents, foremen, subforemen, electricians, apprentice, and helpers.

The efficiency of performance of the workers, in nearly all cases, depends upon the personnel relations of the top-level supervisory personnel (owner, engineers, general superintendent) with the lower-level supervisory personnel (job superintendent and foremen) and with lower-level supervisory personnel and the workmen. There are no shortcuts. A high degree of objectivity and cooperation between these employees is absolutely essential in order to carry out the highly specialized task of conducting an electrical contracting business on a profitable basis.

Data presented in this chapter is designed to acquaint the reader with current labor requirements and practices, labor laws and agencies, handling grievances and disputes, personnel relations, and similar items considered necessary in running the labor and personnel requirements of an electrical contracting business as smoothly as possible. Safety precautions are also be included to understand the latest requirements of OSHA.

10.1 Labor Requirements and Practices

Electrical contracting is a highly specialized business of installing, altering, and maintaining electrical installations for building construction. Therefore, specially trained workmen are required to carry out the day-to-day functions of such a business. Such workmen normally fall into one or more of the following categories:

1. Apprentice electricians or helpers.
2. Skilled labor.
3. Specially skilled labor.
4. Low-level supervisory personnel.

Apprentice electricians or helpers are usually under a training program to teach them how to be first-class electricians after a given period of time, like four years. Their normal duties, depending upon the degree of their training, involve such work as assisting in moving heavy equipment, loading and unloading equipment, pulling wire in conduit, and cutting and threading pipe sections to various specified lengths. This type of work requires a good mental attitude to cheerfully accept directions from the more skilled workers to perform many undesirable chores. For example, many persons may resent always getting all the "bull" work when the skilled labor always get the "easier" job and are being paid more. But these are the requirements of an apprentice electrician or helper. It is somewhat like the "Rat" system at a military college—there's an initiation period.

Skilled labor includes electricians who have passed their journeyman's examination and are acquainted with materials, equipment, tools, and installation procedures of most electrical installations. They must constantly keep abreast of the introduction of new types of electrical materials and techniques.

Such workers should be capable of handling conventional electrical installations with the minimum of direction from others, although this is not always the case. There are several journeymen who, because of their attitude or experience, must constantly rely on direction from others to perform their work. Most journeyman electricians, however, in addition to being an efficient electrician, are capable of understanding the details of electrical construction as well as having a good understanding of the NEC, and can perform the required work from working drawings with little or

Figure 10.1. Job superintendent inspecting ground-mounted floodlight.

no assistance from supervisory personnel. These men are definitely foreman material and most will eventually become a foreman over a crew of workers.

Specially skilled labor involves special ability, experience, and knowledge of highly specialized electrical systems such as complex control circuits or cable splicing. Such men usually carry no spe-

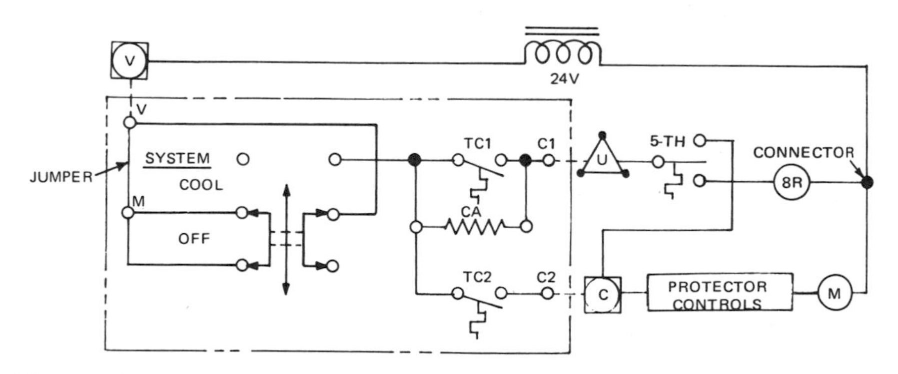

Figure 10.2. Besides being able to guide and direct workers, the job foreman should also have an exceptional knowledge of electrical construction, like reading schematic wiring diagrams.

Figure 10.3. The job foreman should be available to assist inexperienced workers in proper construction techniques.

cial credentials, but their capabilities are soon known throughout the areas in which they work. Such men usually can demand a slightly higher wage than the conventional electricians.

Those workers who show exceptional knowledge of electrical construction and who also have the ability to give directions and guidance to other workers in the performance of the electrical work are usually promoted to higher-level positions such as job foreman and superintendent.

Foremen are usually responsible for the detailed installation supervision which involves the laying out of the work and guiding or "pushing" the workers in performing the installations. When only a few men are under the foreman's supervision, the foreman often is required to work along with the men, but, of course, he is entitled to a higher wage.

Job superintendents are required to handle the complete job; that is, scheduling the installation sequence, scheduling material and tool requirements, selecting foremen and job crews, and maintaining on-the-job cooperation between other trades general contractor and the architect/engineer.

10.2 Labor Laws and Agencies

There are two principal sources of electrical personnel: groups of electrical workers affiliated with the International Brotherhood of Electrical Workers and those without any type of organization. Most of the time the former workers are active members of a local union and are permanent residents of the local area.

There are advantages and disadvantages of each, and each contractor must analyze his own situation to see which road he should take—either union or open shop.

Those contractors who have a working agreement with the I.B.E.W. or local union has the advantage of a "pool" of experienced workers from which to choose. The contractor can therefore call upon this pool at any time necessary to quickly expand his work force and meet sudden job requirements, usually without any trouble. This is especially helpful to contractors whose work varies considerably. When the work load dwindles, the contractor has the option of dismissing any of his employees almost immediately with very little notice. For these advantages, the contractor usually is required to pay a premium on the wages; that is, about 40–60% higher than "open shop" labor.

The "open shop" contractor does not have such a pool to choose from should his work load suddenly require additional workers. Therefore, an "open shop" contractor is required to maintain a steady work force, even between jobs, and try to maintain an even flow of work. The advantages of such a shop are lower wages (in most cases) than organized shops, and less restrictions on the contractor's operation.

Actual experience has proven that there is little difference between the competence of union and nonunion workers on the average, regardless of popular belief.

It is important that the electrical contractor and his supervisory personnel have a good knowledge of labor legislation. Reference should be made to publications available on federal labor laws and agencies. Labor agreements between electrical contractors and electrical workers as represented by a union should also be studied as well as state laws.

10.3 Grievances and Disputes

One of the responsibilities of a job foreman or superintendent is to handle grievances of workers on the job. In the electrical construction industry, a grievance is generally considered to be a localized situation occurring on a specific job due to a justified or unjustified dissatisfaction of one or more workers with the conditions under which he is or may be required to work. The grievance may be due to a claimed violation of management of the terms and conditions of the labor agreement. More often than not, the grievance is based upon a claim of an unfair or unreasonable requirement of the contractor or his supervisory personnel not generally covered by the agreement.

Alleged causes for grievances might include the following:

1. Discrimination in the continual assignment by the foreman of one or more particular workers to unusually hard or distasteful manual labor.
2. Discrimination in the layoff of certain men.
3. Discrimination in the assignment of men for overtime work.
4. Claimed assignment of men to work out of their classification.
5. Claimed speed-up tactics on the part of the contractor or supervisory personnel.
6. Claimed undue restriction of start and stop time, prohibition of coffee time, smoking, and so on.
7. Claimed failure of the contractor or supervisory personnel to provide proper safety equipment or requiring workers to work under hazardous conditions.
8. Claimed attempt of the contractor to force the workers to use "unsafe" types of tools.
9. Claimed violation of union jurisdiction over the assignment or the performance of work by other trades.

In an effort to reduce the possibility of grievances, particularly those that would seriously affect the production on critical or important jobs, supervisory personnel should do the following:

1. Gain the confidence of the workers in assuring them of his interest in handling the work fairly and without discrimination of any worker. This should be by action and not merely lip service. Unnecessary discrimination in the assignment of work tasks, overtime work, premium time work, and so on, should be avoided.
2. Gain the respect of the workers by taking a firm position with respect to compliance with the terms and conditions of the labor agreement

by the contractor as well as the workers. A firm position should be taken with respect to reasonable productive efficiency of the workers on the job.

3. When it is apparent that one or more workers are or may be chronic troublemakers, judiciously transfer them to jobs of lesser importance. Such men may perform excellent work under certain working conditions.
4. Avoid personal controversy with such workers. They will often make every effort to bait supervisory personnel.
5. In extreme cases, such workers should be laid off at the time it is necessary to reduce the work force. In some instances, it may be advisable to discuss the matter with a representative of the union prior to making such a layoff. Doing so will usually preclude any later misunderstanding.
6. Sometimes it may be advisable to discharge such workers at once if it appears that continuing to employ them until the time of reduction of the work force will result in trouble on the job. In such instances, it is advisable to discuss the matter with a representative of the union prior to such dismissal in order to avoid misunderstanding on the job or with the union office.

When the contractor and union representatives on the labor management committee reach an impasse in their attempts to settle a dispute, further recourse is usually possible. The majority of labor agreements include a clause which requires the submission of the dispute to some form of arbitration, usually the Council on Industrial Relations for the Electrical Contracting Industry.

10.4 Personnel Relations

All efforts toward job management will be ineffective if the electricians are unwilling to do their best work or are psychologically incapable of working efficiently. Here is where human relations enter the electrical construction field. Due to the variations of work performed on most electrical construction projects, production-line work is not possible. Therefore electrical installations depend upon the electricians almost entirely.

Every electrical contractor and his supervisory personnel must recognize the fact that every employee is a human being, not a machine that is designed to produce without compensation or compliments. It must further be recognized that the only way to get maximum cooperation and efficiency of performance is not to drive the workers, but to understand them as human beings and make them feel that they are appreciated.

The knowledgeable electrical contractor knows that his men must know what is expected of them if he expects them to take the initiative and perform their work efficiently. He also knows how to get work done *through* people and not *by* people. Finally, the workmen must have confidence in their supervisor before they will cheerfully take directions, instructions, and guidance.

In general, the electrical contractor and his supervisory personnel must treat their electricians as individuals and make use of each person's ability. When a worker is not pulling his load, the supervisor should diplomatically make the worker aware of this. On the other hand, if a job is performed well, the worker should receive a compliment, give credit where credit is due.

Some contractors may have the feeling that that green stuff in the pay envelope is enough compliments and a pink slip means he's not doing his job. However, that worker is human and also is entitled to some personal contact by expression. He should also have the feeling that he has an important role in the overall project.

All employees should know what is "going on" at all times, at least things that directly affect the worker. If any changes are to take place that directly affect the workers (layoffs, transfers, etc.), the worker should be told well in advance if at all possible. When a worker knows what is going on that will affect him directly and indirectly, his attitude and the general morale is improved.

10.5 Safety

The Associated General Contractors of America has published a *Manual of Accident Prevention in Construction* which should be studied by every electrical contractor and his supervisory personnel. This manual has estimated that 50% of all construction accidents could be prevented by using common sense and paying attention to basic safety practice. This manual gives many suggestions how to prevent many of the more common accidents that occur daily in electrical construction work.

Most cities and all states have safety codes and inspection officers to enforce various codes. The codes are available from

Figure 10.4. It is claimed that 50% of all construction accidents could be prevented by using common sense. This opening, for example, should have barricades placed around it rather than just the two boards.

county and state inspection offices. Most areas now utilize the far-reaching, industry-wide standard such as the Occupational Safety and Health Act (OSHA) implemented by the Department of Labor.

As far as the electrical contractor is concerned, OSHA follows the NEC on dealing with electrical installations. However, there are certain requirements that directly affect the electrical contractor and his workers. For example, hard hats are required on most construction projects; all electrical power tools must be provided with a grounded (three-wire) power supply as well as protected with a ground-fault protector; temporary lighting utilizing bare lamps must be provided with approved guards. The electrical contractor must become familiar with all of these activities; there are stiff penalties for not complying.

The principal causes of accidents on electrical construction projects include the following:

1. Failure of equipment and tools; hoisting equipment and slings, failure of ladders, scaffolds, and so on.

2. Improper use of equipment and tools.
3. Attempting to handle excessive weights.
4. Improper installation of materials and equipment.
5. Poor "housekeeping" on jobs.
6. Improper operation of energized equipment and handling of energized circuits.
7. Undue haste.
8. Lack of sufficient personnel.
9. Thoughtlessness.
10. Carelessness.
11. Contempt for observing safety precautions.
12. Lack of knowledge of safety precautions.
13. Fatigue.

11

Completing the Project

After an electrical installation has been substantially completed (all major electrical equipment and components installed and working), there will usually be several areas not quite ready to be turned over to the owner. There are many reasons for this situation: (1) The electrical contractor may have difficulty getting delivery of the last few special lighting fixtures or similar items, (2) the general contractor or one of his subs may be late in installing a piece of equipment—like a heating/cooling unit—which must have electrical connections, or (3) the building site may have to be graded before the electrical contractor can install the outside lighting standards and fixtures.

The owner of the project will be anxious to move in and the electrical contractor should make every effort to get the electrical portion of the job completed at the earliest possible date. Many electrical contractors have installed electrical systems in buildings and kept up with the remaining trades all during the construction phases and then held up the acceptance of the project by not having three or four special battery-operated lighting fixtures. Remember that an electrical contractor's reputation usually is based on the way he finishes a project and not how he begins.

If the hold-up is due to some other trade or something out of the electrical contractor's realm of responsibility, the electrical contractor should still make every effort possible to hurry the other trades along so that the electrical work may be completed. Should such a condition exist, the electrical contractor should make certain that he is not blamed for the situation, even if it

Figure 11.1. Is this project complete? Don't forget the lamps in the chandelier!

means calling a special meeting with the architect and owner. Many people like to "pass the buck" and the building construction industry is no exception.

Near the completion of any project, the electrical contractor should go through the project step-by-step to see if there are going to be any problems with material deliveries or other trades in finishing the project. If any are foreseen, immediate action should be taken. A record will be made of all unfinished areas and the completion of them should be followed through in a methodical manner until the problems are solved and the electrical work is completed.

When the contractor is quite certain that every item is finished, he should call for a final inspection by the architect/engineer, as well as the local municipal electrical inspector (if applicable) as soon as possible. Then if there are any discrepancies—and there are always a few—the electrical contractor will have ample time to correct them without holding up the general contractor or other trades.

11.1 Punch Lists

In some cases, the owner wants to move in a building before all the odds and ends have been completed; he is willing to have small deficiencies corrected later. When such a condition exists, the electrical contractor should make a room-by-room list of all items still missing to save the architect/engineer time. This courtesy will pay off in many ways on future projects under the same architect/engineer.

The list of work to be done should include all items, regardless of how minor, that the contractor detects. For example, a missing switch or receptacle plate, a panelboard cover, a motor connection. Once completed, the project should now be ready for a preliminary inspection by the architect and engineers.

During the inspection, the electrical contractor or one of his superintendents should accompany the inspection team on their tour to be sure that all criticisms are understood. Occasionally an architect or engineer will ask for work beyond the scope of the contract documents, or for items that the contractor may think are beyond the scope of the contract. In this case, the electrical contractor should list all such items so that they may be checked against the contract documents. Then if the contractor feels that the inspection team is wrong, a letter may be written stating the complaint.

On the other hand, if all complaints by the inspection team are justified, as far as the electrical contractor is concerned, a punch list will be forwarded to the electrical contractor within a reasonable time after the inspection. The routing of this punch list usually follows a chain of command; that is, the consulting engineering firm who designed the electrical work on the project will normally send a punch list of all electrical complaints to the architect; the architect in turn will forward a copy of these to the general contractor, along with a list of complaints, deficiencies, and so on for the other trades.

While waiting for the punch list from the general contractor, it would be wise for the electrical contractor to begin work making the corrections since he should already have most of the items if the inspection team was accompanied as recommended earlier. This will save a few days and help the electrical contractor not to delay the completion of the project.

If at all possible, the punch list should be taken care of before the owner begins moving into the building. There are several valid reasons for this suggestion: (1) The owner's workmen will be moving in materials, placing furniture (possibly over or near some punch-list item), and organizing the interior space in general; (2) if any damage is done to the finished electrical work, like breaking a light fixture, after all work has been completed, there is no question as to who pays for the damage. However, if the owner's crew breaks or damages some of the electrical system while the electrical contractor's men are still working, the electrical contractor may end up paying for the damage.

After all the corrections requested in the preliminary punch list have been completed, the electrical contractor should request a final inspection by the local building authorities who will send out one or more electrical inspectors to check the quality of workmanship and to insure that all work complies with the National Electrical Code and all local ordinances.

11.2 Testing the System for Faults

With all the corrections accomplished, the architect/engineer will not have to make another detailed inspection. However, many consulting engineering firms require that a final testing of the electrical system be made in the presence of the electrical engineer or a member from his firm. The extent of this test can vary from job to job and from engineer to engineer, but the following is typical of most electrical specifications:

Testing:

A. The electrical contractor shall take certain voltage and current readings, record all values, and submit in triplicate to the engineer. Two complete sets of readings are required, one under no load and one under maximum available load. The current and voltage shall be recorded on each phase (plus voltage between phases) at main panelboard and at each branch circuit panelboard. Additional spot readings shall be made if required. Resistance of grounding conductors shall be tested and recorded. Forms for submitting this report may be obtained from the engineer's office (see Figure 11.2).

ELECTRICAL TEST DATA REPORT

SOWERS, RODES & WHITESCARVER
CONSULTING ENGINEERS
ROANOKE, VIRGINIA

Project: ____________

Electrical Contractor: ____________
Date tests were made: ____________ Date submitted: ____________

Current characteristics: ______ Volts ______ Phase ______ Wires

Type voltmeter used: ____________ When calibrated: ____________
Type ammeter used: ____________ When calibrated: ____________

Service ground - Resistance in ohms ____________
Resistance test must be made with hand crank, magneto type, megger.

PANEL	Voltage No Load		Voltage Max. Load		Max. Load Amperage			
	Phase to Phase	Phase to Gnd.	Phase to Phase	Phase to Gnd.	Phase A	Phase B	Phase C	Neutral

Figure 11.2. Electrical test data report form.

B. The electrical contractor shall also take voltage and amperage readings on each phase of each motor circuit and each resistance heater circuit installed under the contract, and the same must be recorded as described previously. Also record motor nameplate data, actual motor heater protective device, and all other data necessary for selection of heater device. A typical form is shown in Figure 11-3.

The requirements of the previous specification may seem rigid to those contractors familiar only with residential and small commercial projects; for those involved in industrial installations, perhaps it seems lenient. The requirements of an electrical specification for a typical motel reads as follows:

A. It shall be the responsibility of the electrical contractor to connect the electrical loads to provide minimum phase unbalance throughout the building. The electrical contractor shall operate the building under full heating and other load conditions, with full lighting and provide a record of the amperage per phase for each feeder installed to the main distribution panel.

B. As soon as electric power is available and connected to serve the equipment in the building, and everything is ready for final testing and placing in service, a complete operational test shall be made. The electrical contractor shall furnish all necessary instruments and testing equipment to make all tests, adjustments, and trial operations required to place the system in balanced and satisfactory operational condition. He shall further furnish all necessary assistance and instructions to properly instruct the owner's authorized personnel in the operation and care of the system.

C. Prior to testing the system, the feeders and branch circuits shall be continuous from main feeders to main panels, to subpanels, to outlets, with all breakers and fuses in place. The system shall be tested free from shorts and grounds. Such tests shall be made in the presence of the engineer's representative.

D. No circuits shall be energized without the owner's approval.

11.3 Obtaining Final Payments

Upon delivering the owner or architect with a certification of the electrical inspection approval and after the final inspection by the architect and engineer, a certificate will be issued by the architect authorizing final payment subject to the guarantees. If

MOTOR DATA
FOR

MOTOR NO.	MACHINE DRIVEN	HP	VOLTAGE	PHASE	SERIAL NO.	FRAME	AMPS	RPM	CODE	MOTOR STARTER			
										MAKE	NEMA RATING	COIL NO.	STR SIZE

Figure 11.3. Motor data report and test form.

items are still missing, including forms which have not been completed by the electrical contractor, the architect or engineer may recommend withholding a small amount (usually 15% of the total contract price) to cover these items. At such time as the items are corrected, the retainage will be released to the contractor.

Most electrical contractors are required to guarantee all work performed by him for a period of one year; the guarantee covers all defects caused by materials or workmanship under the electrical contractor's responsibility. This means that if during the 12-month period, any defects should show up due to any defective materials, workmanship, negligence of or want of proper care on the part of the electrical contractor, the electrical contractor is responsible for furnishing any new materials as necessary, repairing the defects, and putting the system in order at his own expense on receipt of notice of such defects from the architect. The contractor is not responsible for defects in the electrical system that are caused by negligence on the part of the owner.

A typical "Schedule of Values and Certificate of Payment" form is shown in Figure 11.4. All such forms may differ slightly, but all will be basically the same. The left-hand column (1, 2, etc.) gives the item number; the second column from the left is for a description of the job breakdown (branch circuit wiring, panelboards, feeders, etc.); the column to follow is for the original value of the work as given in your breakdown estimate to the architect or engineer.

As work is completed, you list the previous amount claimed, the value of work completed on this report, and the total work completed to date (on the portion on the column) and finally the percent of that portion of the work completed to date. These items give the architect/engineer some basis to judge the amount of work completed and to authorize payment to the contractor. If this is the final payment, obviously all items used on the very right-hand column should contain the 100% figure.

11.4 Cleaning Up

Long before the final acceptance of the project, the electrical contractor should have already started moving out certain items. Heavy conduit-bending tools will probably be shipped back to the shop or to another job immediately after all large conduit has been installed; cable-pulling machines, likewise, will have left the job long before the final completion. Perhaps the job shack or trailer was moved shortly after the roof on the project was completed as the interior of the building was more convenient to work from.

Still, there will be several items to consider after the job is completed. The contractor should have one or two of his men walk through the building and pick up all wiring devices, outlet boxes, wire, tools, and any other items that were a result of the electrical contractor's men working on the project. Most of these items will be worthless, but it is a good policy for the electrical contractor to clean up his own mess.

Those items of value should be salvaged for use on another project while the "junk" items should be disposed of by the least expensive manner. Sometimes a junk dealer will gladly take the responsibility of cleaning up after an electrical contractor for the scrap copper, aluminum, and so on. In any event, make certain that all of the items under the electrical contractor's responsibility are removed from the job site, and be careful not to damage any of the finished work in the process. This last statement may seem to be a little overly cautious, but an electrical contractor in Northern Virginia recently managed to knock down six or eight surface-mounted fluorescent lighting fixtures while removing a rolling scaffold from a completed project. The scaffold remained for another day to replace the broken fixtures!

11.5 Using the Completed Project as a Springboard for Other Projects

If the electrical contractor's work managed to satisfy the general contractor, architect, engineer, and owner (this is difficult to do these days) this alone is fine recommendation for another project, but there are many other techniques that can be used to land more projects from the one just completed. Here are a few.

Begin by taking photos of the project: a full shot of the entire building, then portions of the electrical system, especially those that offered a challenge or where a new technique was used. Captions should be provided for these photos and then filed under the job name.

The contractor can then use these photos for advertising in the form of a brochure, perhaps a magazine article for one of the trade journals; the uses are nearly endless.

G. O. Form E & B CO-10 (Rev. 5-71)

SCHEDULE OF VALUES AND CERTIFICATE OF PAYMENT No. ______________

____
Institution or Agency

Building or Project ______________________________ Project No. ____________

Contractor ______________________________

For Period ______________ 19____ To ______________ 19____

Item No.	DESCRIPTION	ORIGINAL SCHEDULE OF VALUES	VALUE OF WORK COMPLETED			
			Previous Value	Value This Report	Total To Date	Percent Complete
1						
2						
3						
4						
5						
6						
7						
8						
9						
10						
11						
12						
13						
14						
15						
16						
17						
18						
19						
20						
21						
22						
23						
24						
25						
26						
27						
28						
29						
30						
31						
32						
33						
34						
35	TOTAL					
36	CHANGE ORDER-PLUS-MINUS					
37	ADJUSTED TOTAL					
38	LESS 10% RETAINED	XXXXXXXXXXXXXX				
39	NET TOTAL	XXXXXXXXXXXXXX				
40	AMT. THIS CERTIFICATION	XXXXXXXXXXXXXX	XXXXXXXXXXXXXX		XXXXXXXXXXXXXX	

Date ______________________ 19____ Contractor ______________________

By ______________________

Figure 11.4. Schedule of values and certificate of payment form.

G. O. Form E & B CO-10 (Rev. 5-71)

SCHEDULE OF VALUES AND CERTIFICATE OF PAYMENT

Page 2

This is to certify that, in accordance with the terms of a contract executed the ______ day of ______ 19 ______ by and between, ______ contractor, and the Commonwealth of Virginia, owner, for work at ______, there is due to the contractor the above-stated amount of ______ $ ______.

Date ______ 19 ______

______
Architectural or Engineering Firm

By ______

Approved:

______, Resident Inspector Date ______ 19 ______

______, Local Build. Comm. Date ______ 19 ______

______, Date ______ 19 ______

LIST CHANGE ORDERS APPROVED DURING THE PERIOD

ORDER NO.	DESCRIPTION	PLUS OR MINUS	AMOUNT	
XXXXXXXX	Previous approvals			
XXXXXXXX	Approved in this period, list below:			
	Total to line 36, page 1			

(List materials on hand at end of period and state value
and any other comments which would be helpful.)

Figure 11.4 (Continued)

Next try to obtain letters of recommendation from the architect, engineer, general contractor, and owner. These may also be used in the contractor's brochure or else used to obtain new projects from other firms who are not familiar with your work or capabilities. One relatively new electrical contractor obtained the contract for a large department store some years ago, and performed his work in the highest manner. He obtained letters of recommendation from all concerned and used these letters to obtain further contracts on all of the chain department store's projects on most of the east coast.

Contacting potential customers directly has been considered the best form of advertising since electrical construction began. With a well designed brochure containing photos and letters of recommendation from previous projects, the prospective customers are going to look twice before giving the contractor the cold shoulder. The brochure should also give personnel qualifications, and a brief history of the organization to promote customer's confidence.

A contractor may wish to hold a party in the completed building after it is completed and invite the owner, architect, engineer, general contractor of the project, and of course other potential customers. Such affairs must be managed carefully, and permission must of course be granted by the owners of the building.

Christmas cards, favors, gadgets, and the like with the electrical contractor's name on them may be worthwhile for spreading one's name around, but an electrical contractor's best advertising is his completed work.

12

Operating Costs and Profit

The operation of any business for profit requires the ability to apply honestly the following:

material cost + labor cost + direct job expense
+ overhead = prime cost
cost + profit = selling price

Because of the highly competitive field of electrical contracting, the electrical contractor must be extremely careful about applying accurate costs and a reasonable profit in calculating the selling price for a given project. If he overestimates his costs, the contractor does not get the job; if he underestimates the job, he gets the job but with too low a profit, or with a loss. It doesn't take too many low-profit or loss-inflicting jobs to force the contractor out of business.

This chapter, therefore, is designed to aid the electrical contractor to correctly calculate direct job costs, indirect job costs, overhead, tool costs, profit, and miscellaneous costs. A knowledge of these items will help to insure the electrical contractor a better chance of survival in this highly competitive field.

12.1 Direct Job Costs

In general, direct job expenses are all those costs, in addition to material and labor, that are chargeable to a given job and which would not have otherwise been incurred had the job not

TRUCK OPERATING RECORD

MONTH OF ____________________ STARTING ____________

ENDING ____________

MAKE __________ LICENSE NO. __________ TOTAL MILES ____________

Day	Hours or Mileage	GASOLINE		OIL		OTHER	
		Gallons	Amount	Quarts	Amount	Item	Amount
1							
2							
3							
4							
5							
6							
7							
8							
9							
10							
11							
12							
13							
14							
15							
16							
17							
18							
19							
20							
21							
22							
23							
24							
25							
26							
27							
28							
29							
30							
31							
TOTAL							

Figure 12.1. One way to keep a record of the company's truck expenses is to use a Truck Operating Record card as shown here.

been undertaken. These costs differ from overhead expenses in that overhead expenses are all of the costs that have to be paid regardless of whether the job was undertaken.

Such items that fall under direct job expenses include permits, fees and special licenses for a given job; withholding taxes and other payroll expenses of productive labor payroll; travel expenses for a given job; postage and freight chargeable to a given job; telephone calls chargeable to a given job; bid bonds and similar items when they can be charged directly to a given project.

In certain cases, the preparation of shop drawings and engineering fees can be charged directly to a job, but are often charged as overhead. This practice may not affect the overall profits of the firm, but it certainly gives a false account of the actual expenses consumed on a given project. This could be detrimental to estimating similar projects in the future.

12.2 Overhead

Overhead expenses are items that cannot be charged to a particular job or project but must be paid to remain in business. Therefore, the contractor's first problem is to determine what his overhead is, and then find some practical method of distributing this amount, on a fair basis, to the various jobs performed over a period of a month, or a year. This is accomplished by adding a certain percentage of the yearly overhead to each contract.

The following items are usually considered as overhead expenses, and a definite amount of money is spent to defray them by all electrical contractors, regardless of their size.

ADVERTISING. Any sales promotion used to obtain or increase business will fall under this heading. Such items as newspaper ads, radio or television commercials, classified phone directory listings, business cards, direct mail brochures, circulars, signs, and truck advertising are included.

ASSOCIATION DUES AND SUBSCRIPTIONS. Business clubs, Chambers of Commerce, electrical associations, and similar dues would be considered under this heading. Also, the subscription to trade journals, if necessary for business operation, may be charged under this heading to qualify as a tax exemption.

AUTOMOBILES AND TRUCKS. All licenses, registration fees, gas, oil, minor repairs, state inspection, and similar costs of automobiles and trucks used in the normal operation of business should be included here.

BAD DEBTS. This account should be debited periodically for uncollectable or doubtful accounts. In planning the overhead, the electrical contractor should make appropriate allowances each year for such bad debts and a proportionate part included in each job estimate.

CHARITABLE CONTRIBUTIONS. All donations to charitable or educational organizations as defined by the federal income-tax regulations should be included under this heading.

COLLECTIONS. The cost of collecting past-due accounts should be charged to this account. This includes legal fees, commissions paid to collection agencies, and so on.

DEFECTIVE, LOST, OBSOLETE, OR STOLEN GOODS. The actual value falling under any or all of these categories is chargeable to overhead and may be deducted from income taxes.

DELIVERY COSTS. The total expense of all freight, express, and postage that cannot be charged directly to a given project belongs in this category.

DEPRECIATION. This account may be debited with the amount of the depreciation on office and shop buildings, autos and trucks, equipment and heavy tools, office furniture, and office equipment, based on the estimated useful life of each.

INSURANCE. Many contractors break this category down into several parts, but insurance fees should be charged as overhead unless a certain percentage of them can be charged as direct job expense. Some types falling under the direct-job-expense category would include fire, theft, and liability.

INTEREST. The amount of interest paid or accrued on outstanding interest-bearing obligations of the business should be charged here.

Figure 12.2. Letterheads and envelopes are a definite overhead expense and can fall under the heading of advertising (when the contractor's name is imprinted) or office expense. In either case, the expense is tax deductible.

LEGAL AND ACCOUNTING FEES. All accountant or CPA fees, as well as those paid to an attorney, fall under this heading.

LOST PRODUCTIVE LABOR. Idle time or time lost in correcting defective work or materials, inventory taking, and similar instances where the electrician's time cannot be charged directly to a job should fall under this category.

OFFICE EXPENSES. Expenses under this heading include such supplies as pencils, pens, staples, printed forms, and similar items, It does not, however, include office equipment like typewriters, calculators, and file cabinets, unless the items are consumed within a year's time.

RENT. The total amount paid for rental or lease of building, land, storage space, and so on used in the operation of the business but that cannot be charged to one particular job would be included here.

REPAIRS AND MAINTENANCE. The total cost of all repairs and maintenance to buildings, equipment, furniture, tools, and so on used in the operation of the business—and that are not capital improvements—are charged to this account.

SALARIES AND COMMISSIONS. This account includes salaries and commissions not readily chargeable to projects. A secretary's salary would normally fall under this heading, but if a secretary was, say, typing written specifications for a particular project, her time could be charged directly to the job.

SALES DISCOUNT. Discounts allowed for either materials or services should be charged to overhead.

SMALL TOOLS. Tools that do not have a prolonged life and are not capital expenditures may be charged under this heading, provided they are not charged to a particular job as direct job expense.

TAXES. Any city, county, state, federal, and other taxes not charged directly to a job should be included under this division. Many contractors break this category down into two parts: general taxes and payroll taxes.

TRAVEL AND TRANSPORTATION. Transportation, meals, and other incidental expenses not directly chargeable to a job but necessary for the operation of the contractor's business should be charged to this account.

TELEPHONE. Charge this account with all telephone calls and services (including answering services) that cannot be charged directly to a particular project.

MISCELLANEOUS. This includes all other expenses not covered previously that are allowed by the IRS and cannot be charged directly to any given job.

The first step in determining the amount of applicable overhead for each job is to determine the base cost, including material cost, labor, and direct cost, to which the overhead percentage rate is to be applied. For new contractors, this percentage will have to be estimated, based on good judgment and from an examination of records or experiences of other contractors. Once the contractor has developed his own records, he can then determine his own overhead more accurately. This determination should be based on several years' operations rather than only a few months'.

The contractor should then determine whether a single rate of overhead can be equitably applied to all jobs or whether, because of both large and small projects under contract, it will be necessary to develop more than one rate. In other words, the overhead rate for smaller projects is more than it would be for larger ones.

The final step is to apportion the total general overhead cost to the overhead cost totals for large and small jobs, respectively, on the basis of percentages used in segregating the rate base for each type of work. The rate base applicable to that type of work is applied accordingly.

12.3 Tool Costs

Many types of tools are used extensively in the installation of electrical systems and the furnishing of such tools by the electrical contractor requires a relatively large investment, not only to make the initial purchase, but also to maintain and repair them. Further expenditures are made to cover insurance on tools as well as interest on the investment and storage of them.

Few electrical contractors are aware of this investment, or at least do not provide some means of recovering their investment. Therefore, the items are charged to overhead. The problem is that very few contractors allow a sufficient overhead charge to always defray the tool and equipment expenses and therefore do not make as much profit as possibly could be made.

One way to cover the expense of tool ownership is to charge a certain percentage to each job. This percentage is based on the initial purchase price of each tool, expense, and use periods. For example, an itemized list of all tools owned by the contractor should be made to which a monthly unit charge should be made. When a job is bid, the estimated length of time certain tools will be used on the project should be determined and then pro rata unit charges should be made to the job.

Since tools can be identified, to a great extent, for each job, they are logically a direct job expense and should be charged as such.

Other methods of recovering tool costs include charging a lump sum per month for the typical number of tool and equipment items provided for a particular type of job requiring a typical number of electricians; a percentage of the entire cost or billing for the job; a percentage of the labor cost of billing or a pro rata charge per man-hour of labor estimated or actually used.

For example, it may be estimated, among other items, that a power threading machine will be used on a job for 6 months. The total initial cost of the machine was, say, $1500 and by referring to "Average Ownership Expense of Tools" schedules (furnished by National Electrical Contractors Association and National Price Service to name two), the contractor can find that the expense of such a machine is 4.5% per working month. Therefore, the contractor should include 4.5% × $1500 = $67.50 per month or $405.00 for the six-month usage.

This figure includes initial purchase price, interest on investment, insurance, replacement costs, and so on.

12.4 Profit

While many electrical contractors derive some pleasure from operating their business, few, if any, would continue their operation of a reasonable profit could not be made. This is the main reason for being in business: to make money.

In general, the amount of profit that an electrical contracting firm should realize depends mainly upon the gross amount of work the firm will perform during a given period of time. For example, if a contractor himself could make $15,000 a year working for another contractor, it stands to reason that he should at least aim for $20,000 annually, considering the headaches of running such a firm. Often, however, the first year or two after starting an electrical contracting business, the profits will be low; but after that time, the contractor should be putting more money into his own pocket than he could make working for someone else.

At first glance, it may seem very simple to operate an electrical contracting business, but the truth is that there are many complications. Items such as constantly fluctuating material prices and labor problems are prime examples. A further complication arises from the fact that the average contractor has difficulty in determining his overhead expenses accurately and more often than not underestimates them. Since his costs are higher than anticipated, his profits are naturally lower.

The business is further complicated, in most cases, by intense competition which makes it compulsory to come up with a correct price on a sufficient amount of work to keep the firm busy.

To obtain the amount of profit necessary to make the business of electrical contracting worthwhile, the contractor must be able to buy materials at the best possible cost, have electricians trained in the type of electrical work anticipated, and practice good job management. These are the basic steps. Then the contractor must know, almost exactly, what his materials are going to cost for any given job; he must apply the correct labor units to these figures; and finally, he must add all direct job expenses to come up with a prime cost for the project.

When the overhead is applied to the prime cost, the contractor should arrive at a price that should cause him to break even. That is, if he took the job calculating all materials, labor, direct job expense, and overhead, he should be able to do the job without losing any money. He will not, of course, make any profit, as this has yet to be added.

Now comes the profit and selling price that can be determined only by the contractor and his staff. Few contractors who perform less than $250,000 worth of electrical construction annually can get by on less than 20% profit and many charge 25–30% profit. On the other hand, the larger contractors often take projects at from 5–10% profit depending upon the type of work. This may not seem like too much profit, but on a project of over $1 million dollars, competition is great and this percentage is just about tops.

In any event, when the contractor has decided upon the percent profit he desires to make on a particular project, this should be added to the prime cost. The result (hopefully) is the selling price that will yield the profit that the contractor has anticipated.

Whether you are a budding electrical contractor or a veteran in the business, you should keep detailed records of each job, giving a cost breakdown on the various stages of completion. Then, if you're making a mistake in your estimates, you should be able to detect the problem immediately and rectify it. This way, you may lose some money on one or two jobs, but if you know your problems and correct them, you can make this money up on future projects. Any losses that occur should be charged against overhead.

13

Cost Accounting

One weakness of those entering a new business is their failure to recognize the importance and need of good accounting records and related bookkeeping procedures. Hip-pocket bookkeeping systems are common in one-man operations, and this is one of the reasons why so many new businesses fail within the first year or two.

Every electrical contractor needs some system to be able to bill customers for work performed (keep track of what money is received and what is still due), keep track of the amount due suppliers and others, make up payroll, and maintain required records and other general operating expenses. Another important reason for bookkeeping systems is to provide a basis for preparing federal and state income tax reports.

The objective of this chapter is to emphasize the need and benefits of proper accounting records and to provide some guidelines to follow in selecting the most applicable bookkeeping system for each contractor's individual needs. Details of bookkeeping and accounting procedures are not provided, but sufficient information is included to aid the contractor in management decisions.

13.1 Need for Good Records

The benefits of good accounting records for an electrical contracting firm are many, but they can be summarized as follows:

1. Records show what has taken place in the past and form a basis for determining why. They then give the electrical contractor a look at his past, both good and bad, and give him a chance to correct the past mistakes as well as indicating the direction for better profit and business opportunities.
2. Good records enable the electrical contractor to compare his operations with industry statistics and trends.
3. Good records present an orderly and complete picture of the firm's financial position which lets the contractor know exactly where he stands. This also helps in obtaining credit from suppliers and banks.
4. The correct use of good records facilitates the preparation of payroll and tax returns.
5. Adequate records can also help in the collection of accounts when invoices are disputed. On the other hand, they can prevent the contractor from having to pay bills that are not due.

13.2 Accounting Records

The individual records of all accounts used in the electrical contracting business should be kept in a book called the general ledger. In most cases, a separate page is kept for each account. All transactions occurring in the business are ultimately summarized in this ledger.

It is also customary to make a record of daily transactions in the order they occur in a general journal. For each entry, a corresponding posting must be made to the general ledger. During an accounting period, there will be as many postings to the various accounts and if a single group of transactions can be combined in a single posting, much time and effort can be saved.

In order to simplify the bookkeeping operation, records called special journals are used to record all transactions of a similar nature. The transactions recurring most frequently in a contracting business are sales, purchases, cash receipts, cash disbursements, payroll, and a brief description of each follows.

SALES JOURNAL. This journal is used to record all sales on account of a major nature. All minor sales are entered in the general journal and all cash sales are entered in a cash receipts journal.

PURCHASE JOURNAL. During the normal course of business, many purchase transactions take place during the month, and a purchase journal is normally utilized to list all purchases on account. Cash purchases, on the other hand, are entered in the cash disbursements journal.

CASH RECEIPTS JOURNAL. The cash receipts journal is utilized to record all transactions regarding monies received in either cash or check.

CASH DISBURSEMENTS JOURNAL. Funds are disbursed to accounts owed by the electrical contractor, for expense charges, and other miscellaneous purchases. All of these disbursements are kept in the cash disbursement journal.

PAYROLL JOURNAL. The cash disbursement journal is designated as the means of recording all cash pay-outs, but most contractors find it more practical to use a special payroll journal for the payroll. Totals are posted directly from this record to the accounts affected in the general ledger. The payroll journal makes it much easier to keep track of payroll withholdings.

ACCOUNTS RECEIVABLE LEDGER. Since electrical contractors, especially those involved in service work, have numerous accounts receivable during the period of a year, it is advisable to keep the accounts in a separate ledger called the accounts receivable ledger. All entries for sales or services as well as payments on these accounts should be made. Any returns, allowances, or discounts should also be noted.

ACCOUNTS PAYABLE LEDGER. This ledger is used to keep a separate record of each creditor who has granted credit to the business. Typical entries will include purchase of materials, tools, and so on, acquiring outside services, payments on account, discounts taken, and all returns and allowances.

JOB COST LEDGER. This is one of the most important records a contractor can keep. It provides a means to compare actual costs of projects with those estimated and will help the contractor avoid any previous mistakes in the future.

13.3 Choice of System

An accounting system integrates the various forms, discussed previously, into a functional operating system. The system provides for the orderly entry of the figures onto the accounting records and an orderly basis of securing from those records the needed or desired information and end products in the form of customer invoices, job cost records, financial statements, and so on. Different bookkeeping or accounting systems vary considerably in detail and method although all follow the same basic principle. The system that will best fit your need will depend upon the type and volume of work performed as well as your financial capabilities.

Of the two fundamental bookkeeping methods, most electrical contractors find the double entry system the best for their needs. This system provides records which cover all activities of the business as well as furnishing a means for recording the effect of each business transaction.

For the new contractor starting out with a basic crew of electricians to engage primarily in small commercial and residential construction, a manual or hand method of bookkeeping will suffice. A complete system is available for less than $25 from New England Business Service, Inc., Townsend, Massachusetts 01470. This system is designed to enable those with little or no experience in bookkeeping to interpret and analyze the progress of their business including the uncovering of hidden sources of profit and loss. The combination of journals, ledgers, payroll records, and so on in one binder adapts to the bookkeeping requirements of any small electrical contracting business.

To set this system up, merely fill in the master header form only; these titles will then be visible above columns on other pages, making it unnecessary to write column headings on every page in the binder.

The system comes in a complete set and extra sheets are available individually when needed. The system includes the following items:

Multi-ring binder
Title sheet
Instruction sheet
Divider sheets with plastic tabs
Divider sheet with untitled tab
Header sheets for income section
Cut journal sheets for income section
Header and cut journal sheets for the disbursement section
Annual payroll summary sheet
Individual payroll record sheets

Annual account summary title sheet
Annual account summary sheets
Trial balance work sheets
Journal adjustment sheet
Equipment register sheet
Accounts receivable control sheet
Header sheets

Once the contractor begins to pick up several accounts, he may want to change to a mechanical bookkeeping system which involves the use of a so-called bookkeeping machine and a reproducing machine. This system, of course, is going to cost much more than the manual system, so the volume of business must be such to warrant such a purchase.

Many of the large electrical contractors have now gone to the electronic data processing bookkeeping systems which processes data through the use of punch cards or punch tape equipment. The tabulating of the information punched on the cards or tape may be either by the use of a professional tabulating service or by tabulating equipment purchased by the contractor.

The ultimate in bookkeeping systems, of course, is specially programmed computer systems related to electrical construction. These systems are becoming more and more popular each year with electrical contracting firms. The service is usually leased from one of the many agencies throughout the country. A central computer is located at the main headquarters of the agency and the contractor rents time for its use. The basic cost for such a service is approximately $300 per month. The contractor then pays for the time it takes to handle a particular requirement. In this way, the contractor can take advantage of the most powerful computers available, without making any capital investment or hiring additional personnel.

One of the principal advantages of the computer is that the contractor obtains more information faster than with other systems. When proper information is fed into the computer on a daily basis, the contractor further has the advantage of obtaining the cost to date on any project in only seconds. This enables the contractor to keep a close watch on all projects and if one is in trouble (going over the alloted budget), the contractor can take immediate steps to correct the problem.

As mentioned previously, most contractors are unfamiliar with bookkeeping and most must therefore rely on outside consultants

(CPA, accountant, etc.) for recommendations or else look to manufacturers of various systems in order to decide which system would best fit the individual contractor's needs. However, when taking the latter route, remember that manufacturers are going to try to sell their system or product, so the contractor should investigate several systems before deciding on any one system.

One of the best ways to determine if a new bookkeeping system is necessary is to ask yourself if your present system is completely satisfactory; that is, does your present system give you all the information desired at any given time? If a contractor is dissatisfied with his present system, he definitely should investigate other systems because it is probably time for a change.

The following table is designed to aid the contractor in selecting the type of system that will best suit his own needs. The table lists the bookkeeping method, the minimum feasible and practical maximum entries per month.

Type of System	Entries per Month	
	Minimum	Practical max.
Manual	Any	1,500
Mechanical	250	5,000
Electronic—punch card	500	6,000
Computer	2000	unlimited

Obviously, this chapter has not given minute details relative to bookkeeping and accounting procedures as used in the electrical contracting business. Rather, it was designed to provide a contractor with general information as a guide to making management decisions about bookkeeping problem situations.

14

Electrical Drawings

The accelerated demand for the use of electricity in buildings has brought with it more complex electrical systems and the need for greater numbers of practical workmen, technicians, and electrical contractors, all of whom must know how to interpret electrical drawings, wiring diagrams, and other supplementary information found in working drawings and written specifications. This chapter is designed to provide the reader with a review of electrical blueprint reading as it is related to the electrical construction field.

14.1 Types of Electrical Drawings

The construction documents supplied by an architectural firm for a new building usually include all architectural drawings which show the design and building construction details; these include floor plan layouts, vertical elevations of all building exteriors, various cross sections of the building, and other details of construction. While the number of such drawings vary from job to job, depending upon its size and complexity, the drawings almost always fall into five general groups.

1. *Site work:* Site plans include the location of the building on the property and show the location and routing of all outside utilities (water, gas, electricity, sewer, etc.) which will serve the building as well as other points of usage within established property lines. Topography lines are sometimes included on the site plan also, especially when the building site is on a slope.

2. *Architectural:* These drawings normally include: elevations of all exterior faces of the building; floor plans showing walls, doors, windows, and partitions on each floor; and sufficient cross sections to indicate clearly the various floor levels and details of the foundation, walls, floors, ceilings, and roof construction. Large-scale detail drawings may also be included.
3. *Structural:* Structural drawings are usually included for reinforced-concrete and structural-steel buildings. These drawings are usually prepared by structural consulting engineers.
4. *Mechanical:* The mechanical drawings cover the complete design and layout of the plumbing, piping, heating, ventilating, and air-conditioning systems and related mechanical construction. Electrical-control wiring diagrams for the heating and cooling systems are often included on the mechanical drawings also.
5. *Electrical:* The electrical drawings cover the complete design and layout of the electrical wiring for lighting, power, signals and communications, special electrical systems, and related electrical work. These drawings sometimes include a site plan showing the location of the building on the property and the interconnecting electrical systems; floor plans showing the location of power outlets, lighting fixtures, panelboards, and so on; power-riser diagrams; and larger-scale details where necessary.

In order to be able to "read" any of these drawings, one must first become familiar with the meaning of the various symbols, lines, and abbreviations used on the drawings and learn how to interpret the message conveyed by each one.

The types of electrical drawings are:

1. Electrical construction drawings.
2. Single-line block diagrams.
3. Schematic wiring diagrams.

Electrical construction drawings show the physical arrangement and views of specific electrical equipment. These drawings give all the plan views, elevation views, and other details necessary to construct the installation. For example, Figure 14.1 shows a pictorial sketch of a wire trough (auxiliary gutter). One side of the trough is labeled "top," one labeled "front," and another labeled "end."

This same trough is represented in another form in Figure

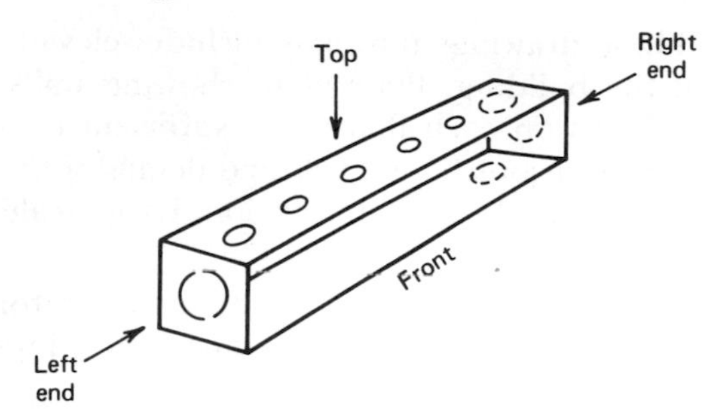

Figure 14.1. Pictorial drawing of a wire trough with top, front, and ends labeled.

14.2. The drawing labeled "top" is what one sees when viewing the panelboard directly from above; the one labeled "end" is viewed from the side, and the drawing labeled "front" shows what the panelboard looks like when viewing the panel directly from the front of it.

The width of the trough is shown by the horizontal lines of the top view and the horizontal lines of the front view. The height is shown by the vertical lines of both the front and the end views, while the depth is shown by the vertical lines of the top views and the horizontal lines of the side view.

The three drawings in Figure 14.2 clearly give the shape of the wire trough, but the drawings alone would not enable a worker to construct it because there is no indication of the size of the trough. There are two common methods to indicate the actual length, width, and height of the wire trough. One is to draw all of the

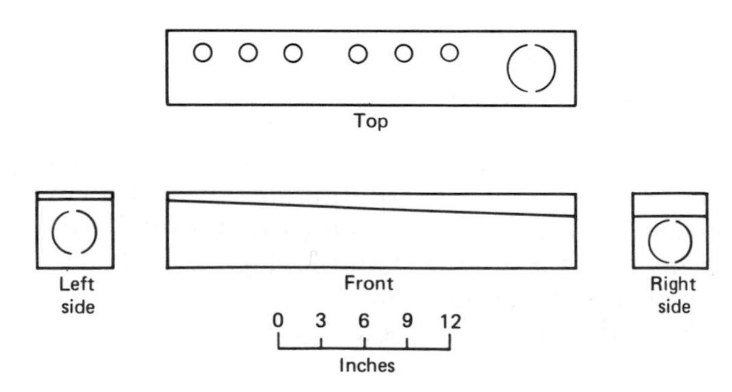

Figure 14.2. Top, front, and side view of the wire trough in Figure 14.1.

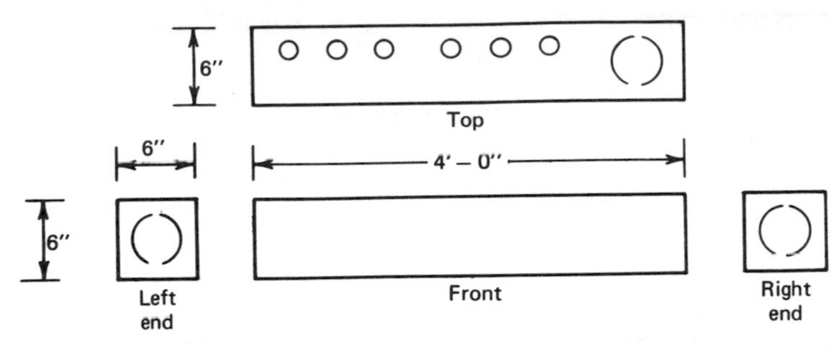

Figure 14.3. Same drawing as Figure 14.2 with alternate method of showing dimensions.

views to some given scale, such as 1½ inches = 1 foot–0 inches. This means that 1½ inches on the drawing represents 1 foot in the actual construction of the housing. The second method is to give dimensions on the drawings like the one shown in Figure 14.3. Note that the gauge and type of material are also given in this drawing—enough data to clearly show how the panelboard housing is to be constructed.

Electrical construction drawings like the ones just covered are used mainly by electrical-equipment manufacturers. The electrical contractor will more often run across electrical construction drawings like the one shown in Figure 14.4. This type of construction drawing is normally used to supplement building electrical-system drawings for a special installation and is often referred to as an *electrical detail drawing.*

Electrical diagrams intend to show, in diagrammatic form, electrical components and their related connections. In diagrams, electrical symbols are used extensively to present the various components. Lines are used to connect these symbols, indicating the size, type, and number of wires that are necessary to complete the electrical circuit.

The electrical contractor will often come into contact with *single-line block* diagrams. These are used extensively by consulting engineering firms to indicate the arrangement of electrical services on electrical working drawings. The *power-riser* diagram in Figure 14.5, for example, is typical of such drawings. This particular drawing shows all of the panelboards and related equipment, as well as the connecting lines to indicate the circuits and feeders. Notes are used to identify each piece of equipment, indicate the size of conduit necessary for each circuit or feeder, and the

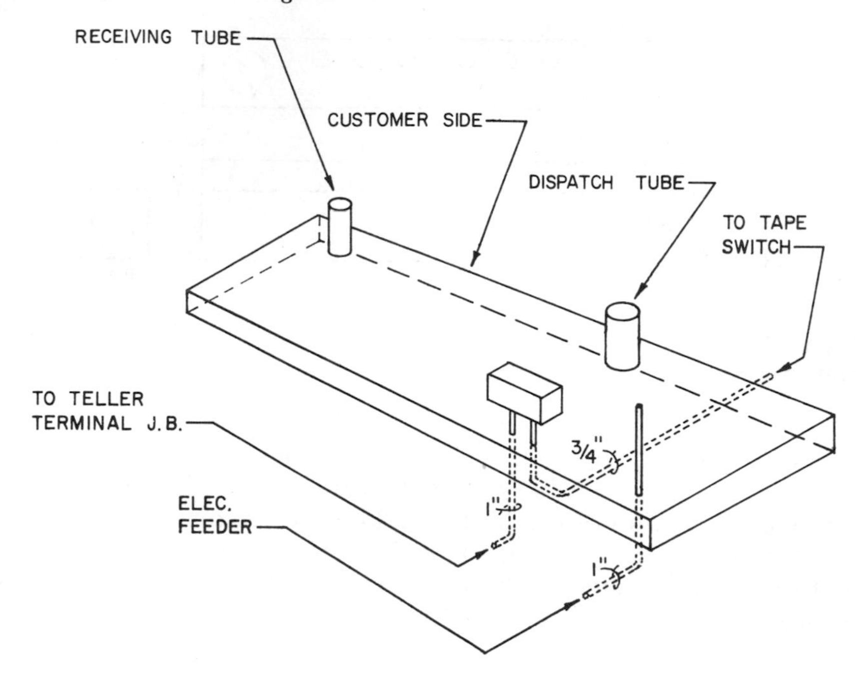

Figure 14.4. An electrical detail drawing used to supplement normal working drawings.

number, size, and type of insulation on the conductors in each conduit.

A *schematic wiring* diagram (Figure 14.6) is similar to a single-line block diagram except that the schematic diagram gives more-detailed information and shows the actual size and number of wires used for the electrical connections.

Anyone involved in the electrical construction industry, in any capacity, frequently encounters all the above (three) types of electrical drawings. Therefore, it is very important for all involved in this industry to fully understand electrical drawings, wiring diagrams, and other supplementary information found in working drawings and in written specifications.

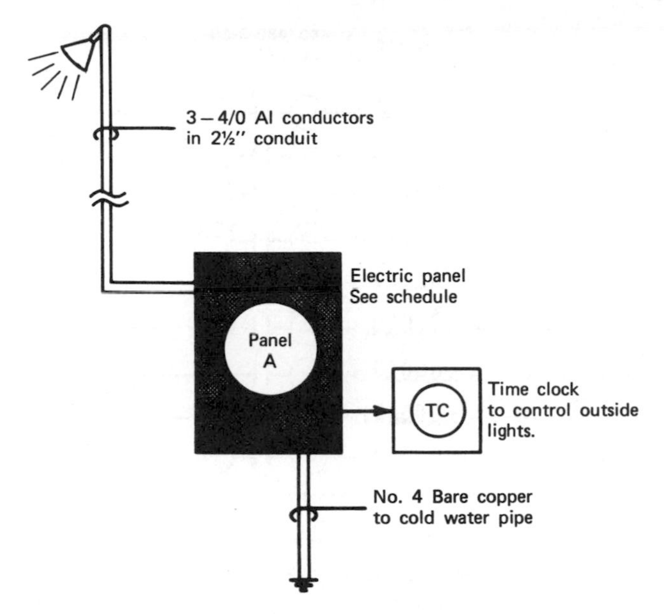

Figure 14.5. A typical power-riser diagram.

14.2 Layout of Electrical Drawings

The ideal electrical drawing should show in a clear, concise manner exactly what is required of the workmen. The amount of data shown on such a drawing should be sufficient, but not overdone. Unfortunately, this is not always the case. The quality of electrical drawings vary from excellent, complete and practical, to just the opposite. In some cases, the design may be so incomplete that the electrical contractor or his estimator will have to supplement it—if not completely design it—prior to estimating the cost of the project or begin installing the electrical system.

In general, a good electrical drawing should contain floor plans of each floor of the building—one for lighting and one for power; riser diagrams to diagrammatically show the service equipment, feeders, and communication equipment; schedules to indicate the components of the service equipment, lighting fixtures, and similar equipment; and large-scale detailed drawings for special or unusual portions of the installation. A legend or electrical symbol list should also be provided on the drawings in order to explain the meaning of every symbol, line, and so forth used on the

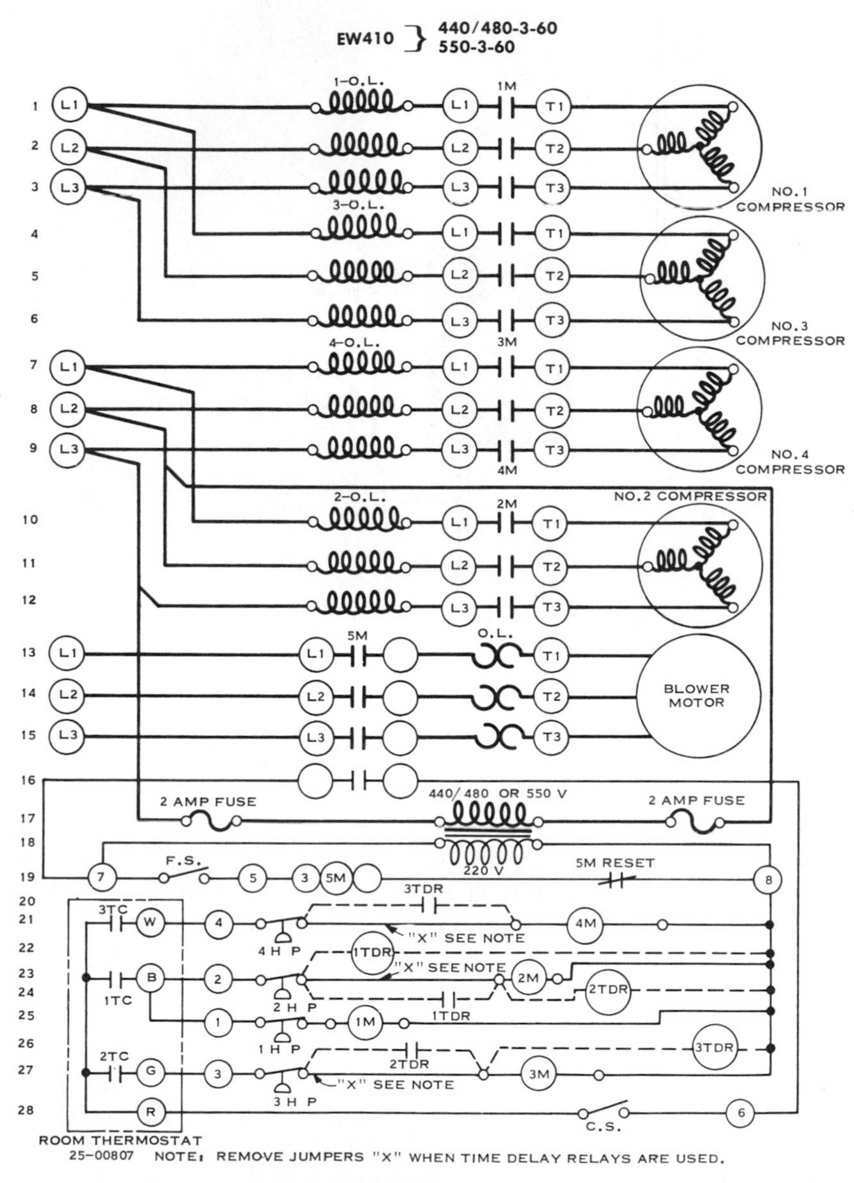

Figure 14.6. A schematic wiring diagram, as shown here, gives more detailed information and shows the actual size and number of wires used for the electrical connections.

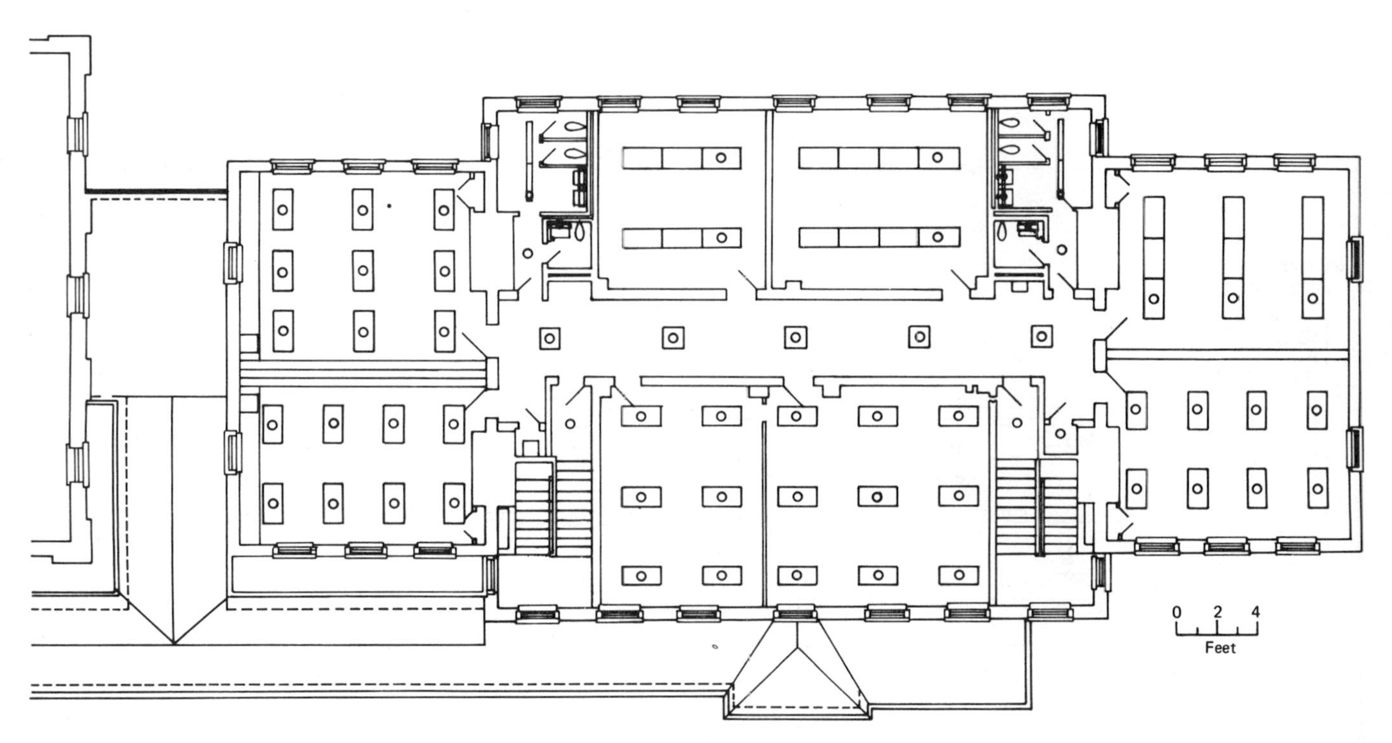

Figure 14.7. An example of a poor quality drawing requiring additional design work prior to estimating or installing the electrical system.

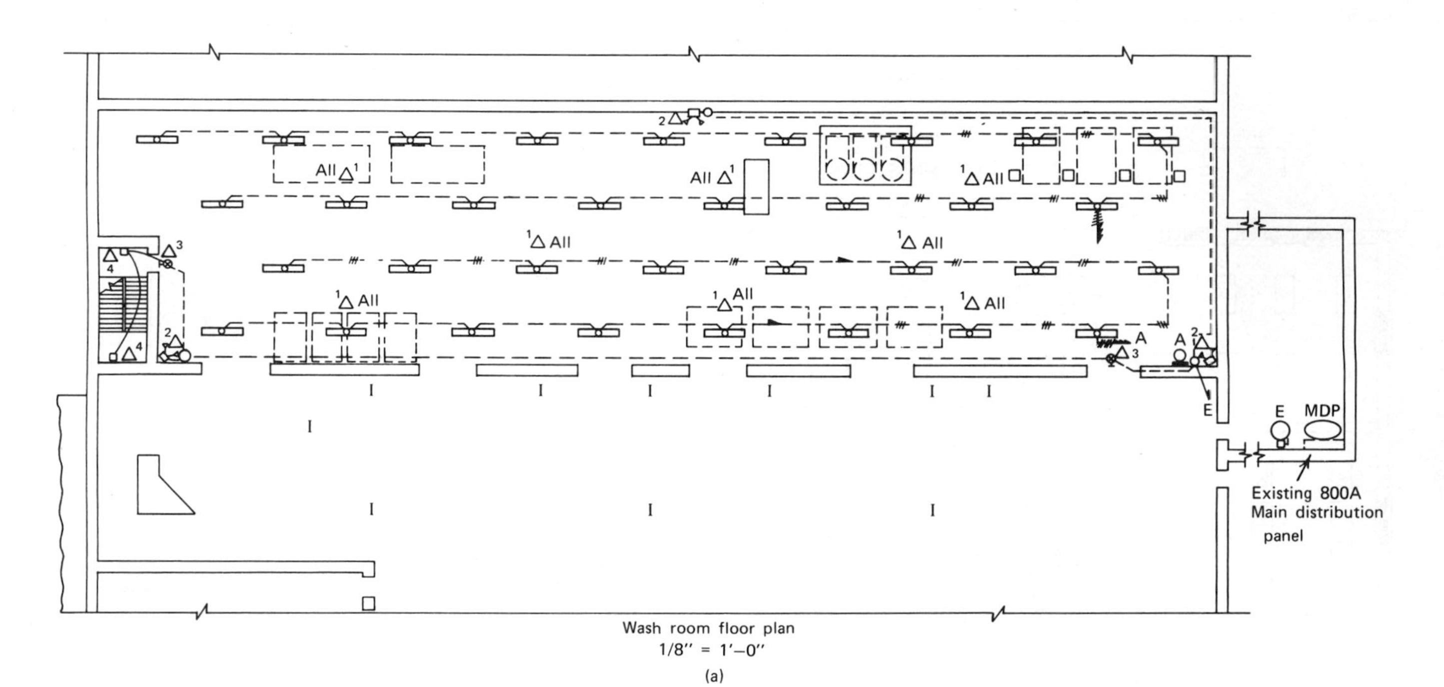

Wash room floor plan
1/8″ = 1′–0″
(a)

Figure 14.8.

ELECTRICAL SYMBOL LIST

- Wall outlet w/quartz lighting fixture.
- Wall outlet w/exit light connected to emergency circuit.
- Ceiling outlet with fluorescent fixture.
- Emergency lighting – see lighting fixture schedule.
- Branch circuit run exposed in ceiling or wall. Slash marks indicate number of conductors in run. No slash marks indicates two No. 12. conductors.
- Homerun to panelboard. No. of arrowheads indicate number of circuits.
- Indicates type of lighting fixture – see schedule.
- Lighting panel.
- Fusible safety switch.
- (J) Junction box sized according to N.E.C.
- S Single pole switch mounted 50" A.F.F. to ℄ of box.

(b)

PANELBOARD SCHEDULE

PANEL No.	TYPE CABINET	PANEL MAINS AMPS	PANEL MAINS VOLTS	PANEL MAINS PHASE	BRANCHES 1P	BRANCHES 2P	BRANCHES 3P	BRANCHES PROT	BRANCHES FRAME	ITEM FED OR REMARKS
A	SURFACE	125 A	120/240 V	3ϕ, 4W	6			20A		LIGHTING
SQ. "D" TYPE NQO W/MAIN LUGS ONLY					1			20A		120 V WASHER
						1		30A		DRYER
					2			20A		RECEPTS
					10			20A		SPARES

LIGHTING FIXTURE SCHEDULE

FIXT. TYPE	MANUFACTURER'S DESCRIPTION	LAMPS No.	LAMPS TYPE	VOLTS	MOUNTING	REMARKS
1	LITHONIA CAT. No. EU 296 HO	2	110W F	120	SURFACE	
2	CHLORIDE SYSTEMS INC. CAT. No. 655 · A4 · L2	2	SEALED BEAM		BRACKET	
3	CHLORIDE SYSTEMS INC. CAT. No. SPX · S · 1	2	15W I		SURFACE	MT. 1'–0" ABOVE DOOR FRAME
4	CHLORIDE SYSTEMS INC. CAT. No. SPU · 10 · 120	1	QUARTZ			MT. 6'–6" A.F.F. TO ℄ OF BOX
5						

(c)

Figure 14.8. A drawing (on facing page) of relatively good quality, showing all outlets, all branch circuits, feeder and service runs, and other pertinent data.

drawings. Anything that cannot be explained by symbols and lines should be clarified with neatly lettered notes or else explained in the written specifications. The scale to which the drawings are prepared is also important; they should be as large as practical and where dimensions are to be held to extreme accuracy, dimension lines should be added. Figure 14.7 shows a poorly prepared electrical drawing while Figure 14.8 shows one of relatively good quality. In the former drawing, it is obvious that the electrical contractor will have to lay out or design portions of the system before it can be properly estimated or installed.

The following steps are usually necessary in preparing a good set of electrical working drawings and specifications:

1. The engineer or electrical designer meets with the architect and owner to discuss the electrical needs of the building in question and also to discuss various recommendations made by all parties.
2. Once the data in No. 1 above is agreed upon, an outline of the architects floor plan is drawn on tracing paper and then several prints of this floor plan outline are made.
3. The designer or engineer then calculates the required power and lighting requirements for the building and sketches them on the prints.
4. All communication and alarm systems are located on the floor plans along with lighting and power panelboards. Again these are sketched on the prints.
5. Circuit calculations are made to determine wire size and overcurrent protection and then reflected on the drawings.
6. After all the electrical loads in the entire building have been determined, the main electric service and related components (transformers, etc.) are selected and sketched on the prints.
7. Schedules are next in line to identify various pieces of electrical equipment.
8. Wiring diagrams are made to show the workmen how various electrical components are to be connected. An electrical symbol list is also included to identify the symbols used on the drawings.
9. Various large-scale electrical details are included, if necessary, to show exactly what is required of the workmen.
10. Written specifications are then made to give a description of the materials and the installation methods.

If these steps are properly taken in preparing a set of electrical working drawings, they will be detailed and accurate enough for a more rapid and accurate estimate as well as a first-class installation.

14.3 Electrical Graphic Symbols and Schedules

Since electrical drawings must be prepared by electrical draftsman in a given time period so as to stay within the allotted budget, symbols are used to simplify the work. In turn, a knowledge of electrical symbols must also be acquired by anyone who must interpret and work with the drawings.

Most engineers and designers use electrical symbols adopted by the American National Standards Institute (ANSI). However, many of these symbols are frequently modified to suit certain needs for which there is no standard symbol. For this reason, most drawings include a symbol list or legend either as part of the drawings or included in the written specifications.

Figure 14.9 shows a list of electrical symbols prepared and used by one consulting engineering firm. This list represents a good set of electrical symbols in that they are: (1) easy to draw by draftsmen, (2) easily interpreted by workmen, and (3) sufficient for most applications.

It is evident from the preceding list that many symbols have the same basic form, but their meanings differ slightly because of the addition of a line, mark, or abbreviation. Therefore, a good procedure to follow in learning the different electrical symbols is to first learn the basic form and then apply the variations of that form to obtain the different meanings.

Note also that some of the symbols listed contain abbreviations, such as WT for *watertight* and S for *switch.* Others are simplified pictographs, such as Figure 14.10 for a safety switch, or Figure 14.11 for a flush-mounted panelboard and housing.

In some cases, the symbols are combinations of abbreviations and pictographs, such as Figure 14.12 for a nonfusible safety switch.

14.3.1 Electrical Schedules

A schedule, as related to electrical drawings, is a systematic method of presenting notes or lists of equipment on a drawing in

ELECTRICAL SYMBOL LIST

NOTE: THESE ARE STANDARD SYMBOLS AND MAY NOT ALL APPEAR ON THE PROJECT DRAWINGS; HOWEVER, WHEREVER THE SYMBOL ON PROJECT DRAWINGS OCCURS, THE ITEM SHALL BE PROVIDED AND INSTALLED.

	CEILING OUTLET WITH INCANDESCENT FIXTURE
	RECESSED OUTLET WITH INCANDESCENT FIXTURE
	WALL-MOUNTED OUTLET WITH INCANDESCENT FIXTURE
	CEILING OUTLET WITH FLUORESCENT FIXTURE
	WALL-MOUNTED OUTLET WITH FLUORESCENT FIXTURE
	FLUORESCENT FIXTURE MOUNTED UNDER CABINET
	GROUND-MOUNTED UPLIGHT
	POST-MOUNTED INCANDESCENT FIXTURE
	FLOOD LIGHT FIXTURE
	FLUORESCENT STRIP
	EXIT LIGHT, SURFACE OR PENDANT
	EXIT LIGHT, WALL MOUNTED
5	INDICATES TYPE OF LIGHTING FIXTURE — SEE SCHEDULE
S	SINGLE-POLE SWITCH MOUNTED 50" UP TO ℄ OF BOX
S_3	THREE-WAY SWITCH MOUNTED 50" UP TO ℄ OF BOX
S_4	FOUR-WAY SWITCH MOUNTED 50" UP TO ℄ OF BOX
S_2	TWO-POLE SWITCH MOUNTED 50" UP TO ℄ OF BOX
S_L	LOW VOLTAGE SWITCH TO RELAY
S_D	DOOR SWITCH
	DUPLEX RECEPTACLE MOUNTED 18" UP TO CENTER OF BOX
	DUPLEX RECEPTACLE MOUNTED 4" ABOVE COUNTERTOP
	SPLIT-WIRED DUPLEX RECEPTACLE—TOP HALF SWITCHED
1	SPECIAL OUTLET OR CONNECTION—NUMERAL INDICATES TYPE—SEE LEGEND AT END OF SYMBOL LIST
	FLOOR-MOUNTED RECEPTACLE
©	CLOCK HANGER RECEPTACLE
	PUSHBUTTON SWITCH FOR DOOR CHIMES
CH	CHIMES
TV	TV OUTLET MOUNTED 18" UP TO ℄ OF BOX
	TELEPHONE OUTLET
	FUSIBLE SAFETY SWITCH

Figure 14.9. A list of electrical symbols typical of those used on electrical drawings.

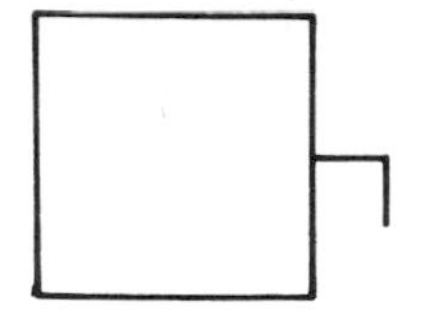

Figure 14.10. Symbol for safety switch.

tabular form. When properly organized and thoroughly understood, schedules are not only powerful timesaving methods for the draftsmen, but also save the electrical contractor and his personnel much valuable time in preparing the estimate and installing the equipment in the field.

For example, the lighting-fixture schedule in Figure 14.13 lists the fixture type from letters or numbers on the drawings. The manufacturer and catalog number of each fixture is included along with the number, size, and type of lamp for each. The "Volts" and "Mounting" columns follow, and the column on the extreme right is for special remarks such as giving the mounting height for a wall-mounted fixture.

Sometimes schedules are omitted on the drawings and the information is contained in the written specifications instead. Besides being time consuming (combing through page after page of written specifications) many workmen do not always have access to the specifications while working, whereas they usually do have access to the working drawings at all times.

The schedules in Figures 14.14 through 14.16 are typical of those used on electrical drawings by consulting engineers.

14.4 Sectional Views

Sometimes the construction of a building is difficult to show with the regular projection views normally used on electrical drawings. If too many broken lines are needed, for example, to show hidden objects in buildings or equipment, the drawings become confusing

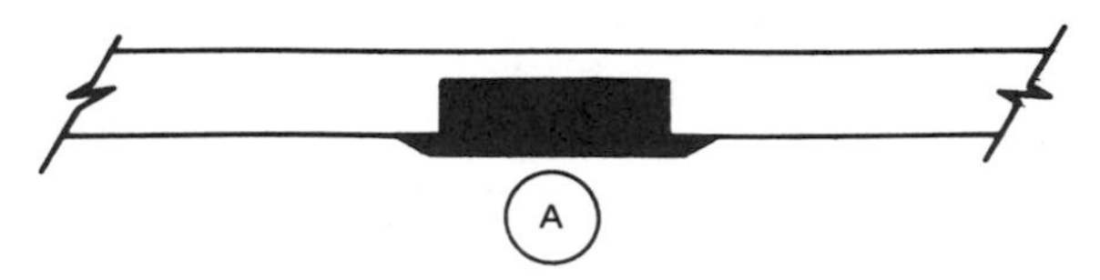

Figure 14.11. Symbol for flush-mounted panelboard and housing.

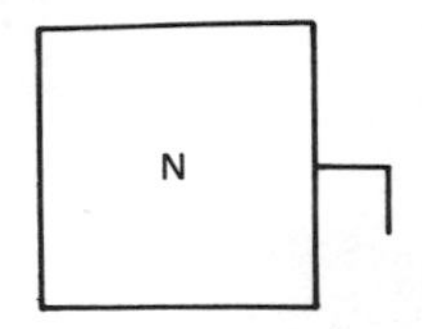

Figure 14.12. Symbol for nonfusible safety switch.

and difficult to read. Therefore, in most cases, building sections are shown on working drawings to clarify the construction. To better understand a building section, imagine that the building has been cut into sections as if with a saw. The floor plan of a building in Figure 14.17 shows a sectional cut at point A-A. This cut is then shown in Figure 14.18.

In dealing with sections, one must use a considerable amount of visualization. Some sections are very easy to read while others are extremely difficult as there are no set rules for determining what a section will look like. For example, a piece of rigid conduit, cut vertically, will have the shape of a rectangle; cut horizontally, it will have a round or circular appearance; cut on the slant, it will be an ellipse.

A cutting plane line (Figure 14.19) has arrowheads to show the direction in which the section is viewed. Letters, such as A-A, B-B, are normally used with the cutting-plane lines to identify the cutting plane and the corresponding sectional views.

14.5 Wiring Diagrams

Complete schematic wiring diagrams are used infrequently on the average set of electrical working drawings (only in highly unique and complicated systems like control circuits), but it is important to have a thorough understanding of them when the need to interpret arises.

Components in schematic wiring diagrams are represented by symbols, and every wire is either shown by itself or included in an assembly of several wires which appear as one line on the drawing. Each wire in the assembly, however, is numbered when it enters and keeps the same number when it emerges to be connected to some electrical component in the system. Figure 14.20 shows a complete schematic wiring diagram for a three-phase, AC magnetic motor starter. Note that this diagram shows the various

LIGHTING FIXTURE SCHEDULE						
FIXT. TYPE	MANUFACTURER'S DESCRIPTION	LAMPS		VOLTS	MOUNTING	REMARKS
		No.	TYPE			

Figure 14.13. Lighting fixture schedule.

FRANK J. SULLIVAN ASSOCIATES
CONSULTING ENGINEERS
WASHINGTON 36, D.C.

PROJECT: ____________ DATE: ____________

JOB: ____________ BY: ____________

INTERCOMMUNICATION SYSTEM SCHEDULE

ROOM NAME	ITEM	I.D. No.	1	2	3	4	5	6	7	8	9	10	11	12	13	14	15	16	17	18	19	20	21	22	23	24	25	26	REMARKS
		1																											
		2																											
		3																											
		4																											
		5																											
		6																											
		7																											
		15																											
		16																											
		17																											
		18																											
		19																											
		20																											
		21																											
		22																											
		23																											
		24																											
		25																											
		26																											

Figure 14.14. Intercommunication system schedule.

LOAD CENTER UNIT SUBSTATION EQUIPMENT SCHEDULE					
ITEM	SWITCH			EQUIPMENT	DESIGNATION
	SWITCH RATING	POLES	FUSE RATING		

Figure 14.15. Load center unit substation equipment schedule.

DATE_ _ _.　　　　PROJECT/JOB NO. _ _ _ _ _ _ _ _ _ _ _ _ _ _ _　　　　MAINS_ _ _A.

BY_ _ _ _.　　　　MOTOR CONTROL CENTER NO._ MCC_　　　　_ϕ_ _ _V.

ITEM NO	HP KW	F.L.A.	STR. SIZE	P'S	SW SIZE	FUSE SIZE	AUX. EQUIP	REM. CONTROL DEVICES	POWER CONTROL DIAGRAM	NAMEPLATE DESIGNATION	REMARKS
											_ _MCC_ _ _

Figure 14.16. Motor control center schedule.

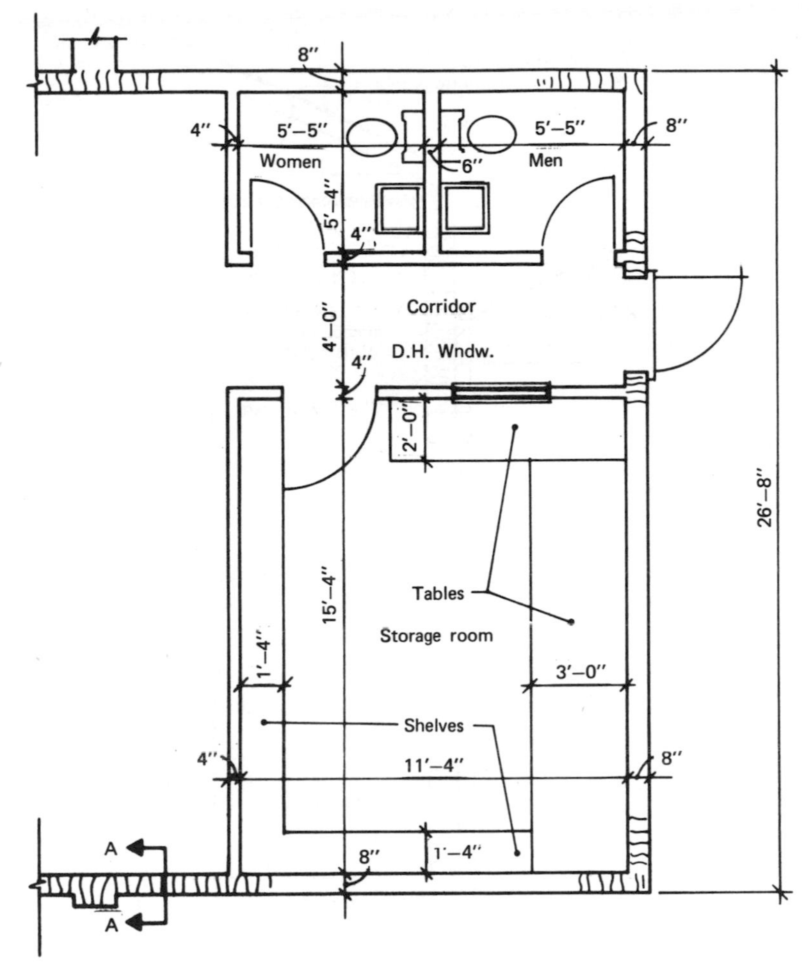

Figure 14.17. Floor plan of a building showing a sectional cut at point A-A.

devices (in symbol form) and indicates the actual connections of all wires between the devices.

Figure 14.21 gives a list of electrical wiring symbols commonly used for *single-line* schematic diagrams; *single-line* diagrams are simplified versions of complete schematic diagrams. Figure 14.22 shows the use of these symbols in a typical single-line diagram of an industrial power-distribution system.

Power-riser diagrams are probably the most frequently encountered diagrams on electrical working drawings for building con-

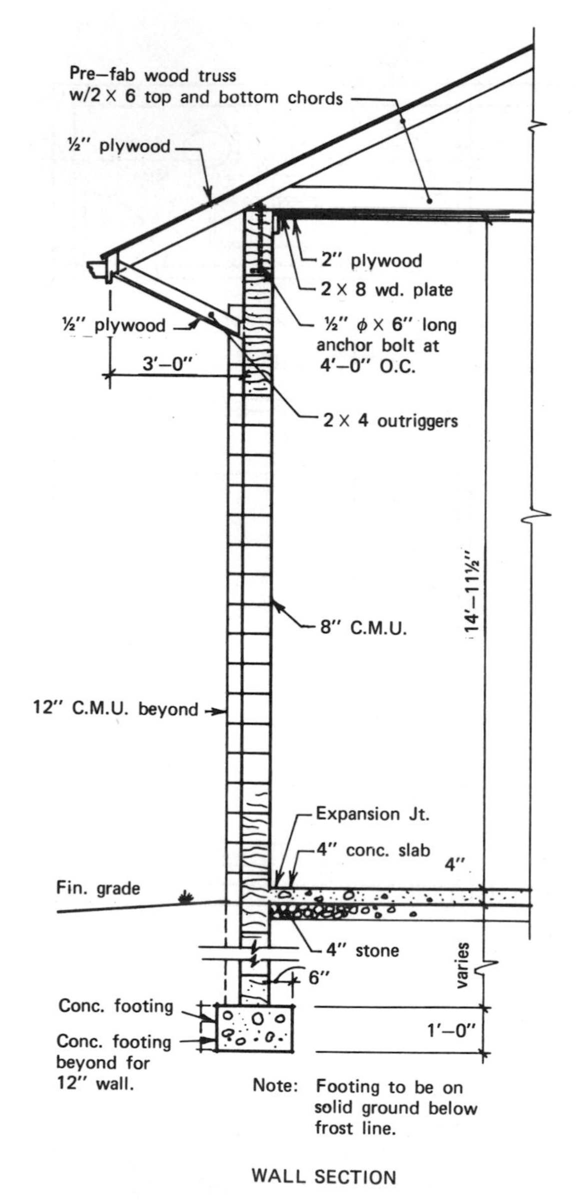

Figure 14.18. Sectional view of the cut in Figure 14.17.

Figure 14.19. Cutting-plane line.

struction. Such diagrams give a picture of what components are to be used and how they are to be connected in relation to one another. This type of diagram is more easily understood at a glance than the previously described types of diagrams. As an example, compare the power-riser diagram in Figure 14.23 with the schematic diagram in Figure 14.24. Both are wiring diagrams of an identical electrical system, but it is easy to see that the drawing in Figure 14.23 is greatly simplified, although a supplemental schedule is required to give all necessary data required to construct the system. Such diagrams are also frequently used on telephone, television, alarms, and similar systems.

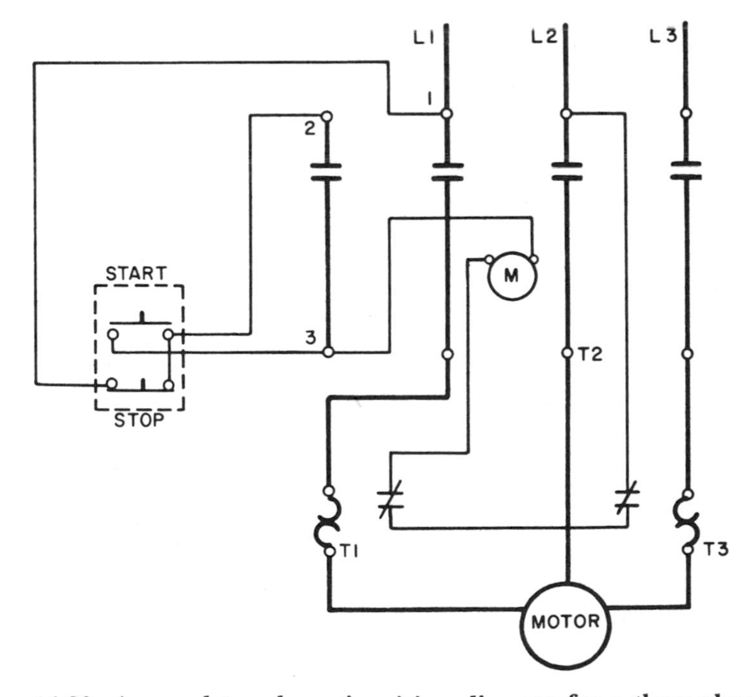

Figure 14.20. A complete schematic wiring diagram for a three-phase AC magnetic motor starter.

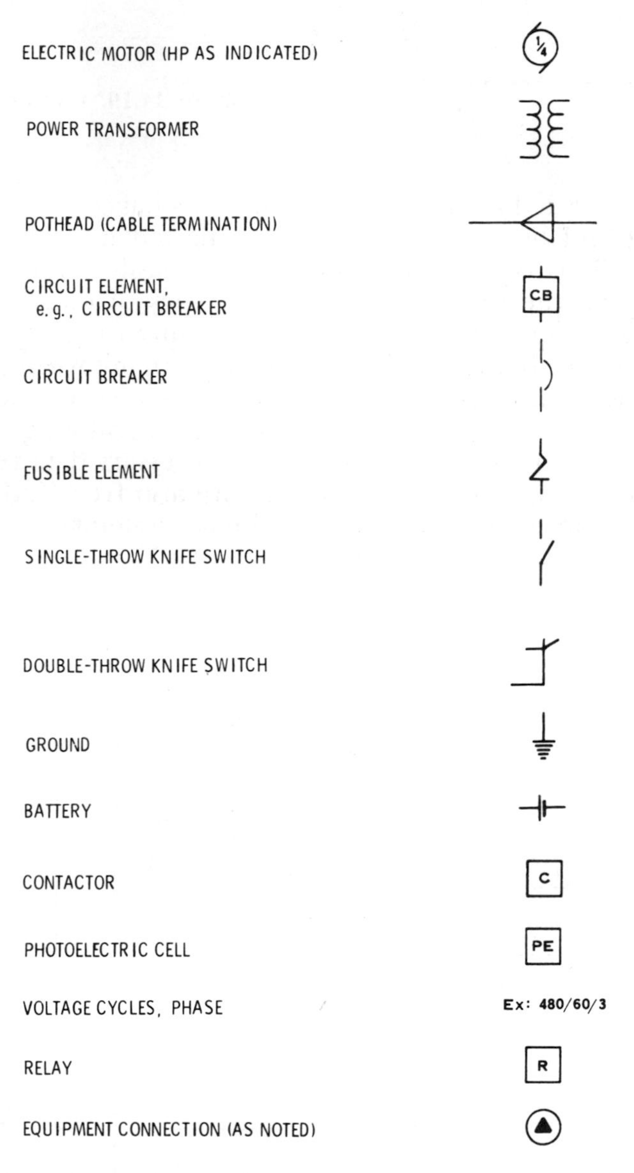

Figure 14.21. A list of electrical wiring symbols commonly used for single-line schematic diagrams.

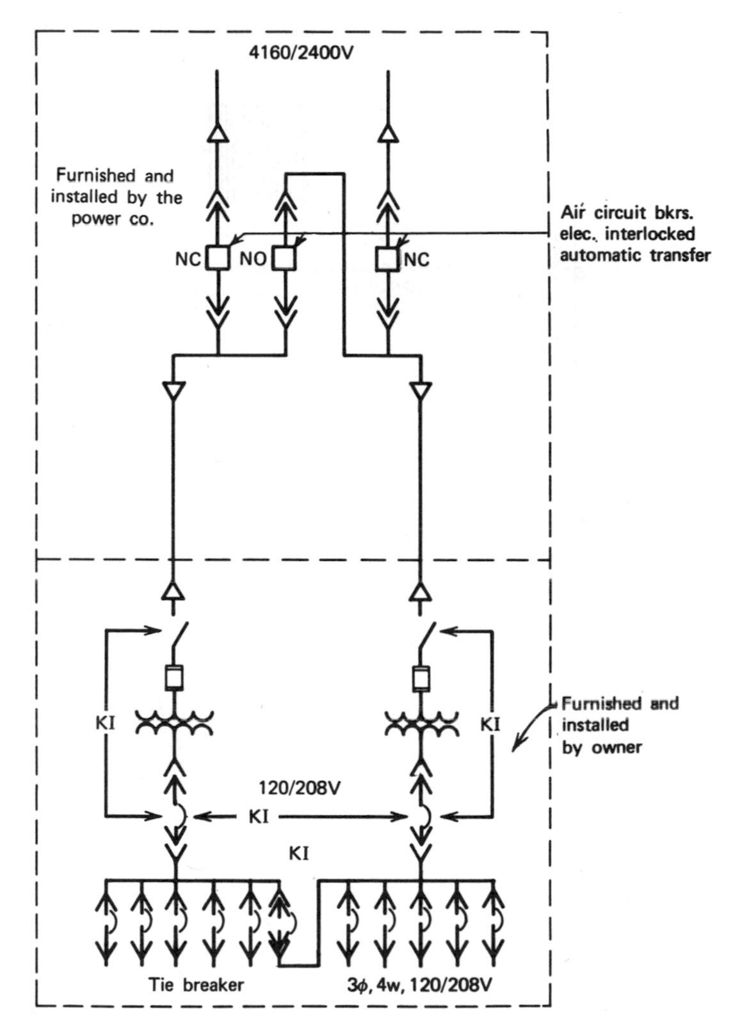

Figure 14.22. A typical single-line diagram using some of the symbols in Figure 14.21.

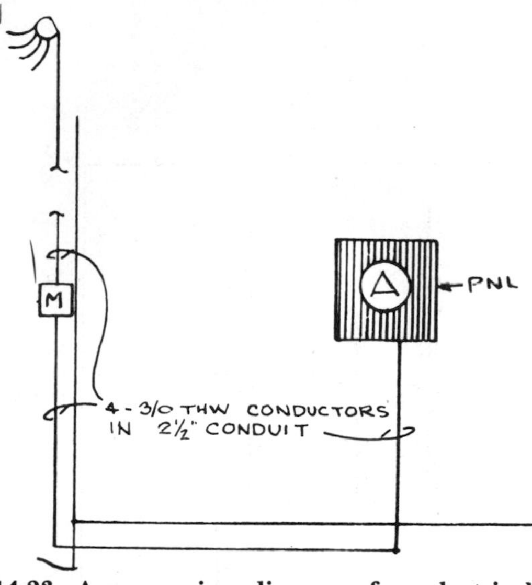

Figure 14.23. A power-riser diagram of an electrical service.

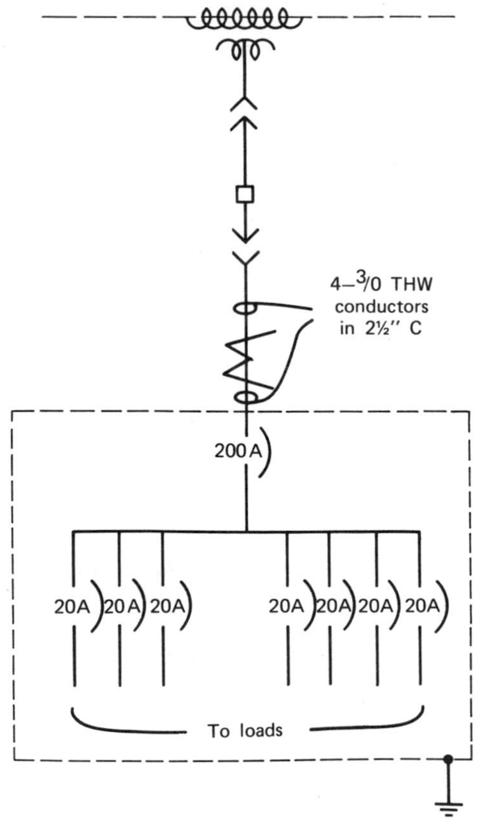

Figure 14.24. A schematic diagram of the electrical service shown in Figure 14.23.

14.6 Site Plans

A site plan is a plan view that shows the entire property with the buildings drawn in their proper location on the plot. Such plans sometimes include sidewalks, driveways, streets, and utility systems related to the building or project.

Site plans are drawn to scale using the engineer's scale rather than the architect's scale used for most building plans. Usually, on small lots, a scale of, say, 1 inch = 10 feet or 1 inch = 20 feet is used. This means that 1 inch (actual measurement on the drawing) is equal to 10 feet, 20 feet, and so on, on the land itself.

In general building-construction practice, it is usually the owner's responsibility to furnish the architect with property and topographic surveys, which are made by a certified land surveyor or civil engineer. These surveys will show: (1) all property lines, (2) existing utilities and their location on or near the property, (3) the direction of the land slope, and (4) the condition of the land (woody, swampy, etc.)

The architect and engineer then use this site plan to incorporate all new utilities. The electrical contractor will then be concerned with the electrical distribution lines, the telephone lines, and possibly cable television lines, especially if they are to be installed underground.

15

Electrical Specifications

The electrical specifications for a building or project are the written descriptions of work and duties required of the architect, engineer, or owner. Together with the working drawings, these specifications form the basis of the contract requirements for the construction of the electrical system for the building or project.

15.1 Sections of the Specifications

Divisions 1 through 16 of the written specifications cover requirements of a specific part of the construction work on the project. Included in these divisions are the type and grade of materials to be used, equipment to be furnished, and the manner in which it is to be installed. Each division will indicate the extent of the work covered and should be so written as to leave absolutely no doubt in anyone's mind of whether a certain part of the work is included in one section of the specifications—to be performed by a certain contractor—or another.

The following is an outline of the various sections normally included in a complete set of construction specifications.

DIVISION 1—GENERAL REQUIREMENTS. This division covers a summary of the work, alternatives project meeting, submittals, quality control, temporary facilities and controls, products, and the project closeout. Every responsible person involved with the project should become familiar with this division.

DIVISION 2—SITE WORK. Usually the only part of this division that would concern electrical workers is "Site Utilities"; that is, electrical, telephone, and perhaps cable television distribution systems running, either overhead or underground, outside of the building.

DIVISION 3—CONCRETE. Work covered under this division includes concrete formwork, expansion and contraction joints, concrete reinforcement, cast-in-place concrete, specially finished concrete, specially placed concrete, precast concrete, and cementitious decks. If the concrete documents call for concrete pads for transformers and other electrical equipment, the requirements will normally be given in Division 16. However, the electrical contractor should refer to this division of the specifications also when such a condition exists.

DIVISION 4—MASONRY. Covers mortar, masonry accessories, unit masonry, stone, masonry restoration and cleaning, as well as refactories. Very little work under this division will involve the electrical contractor.

DIVISION 5—METALS. Structural metal framing, metal joists, metal decking, lightgage framing, metal fabrications, ornamental metal, and expansion control normally fall under this division of the construction specifications. Items concerning the mounting of outlet boxes, conduits, and so on should concern the electrical contractor; that is, the type of clamps and other mounting devices to connect the electrical items to the metal structure.

DIVISION 6—CARPENTRY. Most items pertaining to wood fall under this division; rough carpentry, heavy timber construction, trestles, prefabricated structural wood, finish carpentry, wood treatment, architectural woodwork, and the like. Plastic fabrications are also included if used on the project for which the specifications are written.

DIVISION 7—THERMAL AND MOISTURE PROTECTION. The description of items in this division cover such items as waterproofing, dampproofing, building insulation, shingles and roofing tiles, preformed roofing and siding, membrance roofing, sheet metal work, wall flashing, roof accessories, and sealants. The electrical contractor may want to refer to this section should he be responsible for waterproofing pipe or conduit flanges positioned on the roof of the building; a service mast would be one application.

DIVISION 8—DOORS AND WINDOWS. Few if any of the items in this division should concern the electrical contractor. However, for reference, he may want to know some of the items covered: metal doors and frames, wood and plastic doors, special doors, entrances and storefronts, metal windows, wood and plastic windows, special windows, hardware and specialties, glazing, and window wall/curtin walls.

DIVISION 9—FINISHES. It is a good idea for the electrical contractor to look over the items in Division 9 of the construction specifications. The types of finishes will give the contractor, his estimator, purchasing agent, and so on an idea of the types of outlet boxes required for the various finishes within the building. This division gives the types, quality, and workmanship of lath and plaster, gypsum wallboard, tile, terrazzo, acoustical treatment, ceiling suspension systems, wood flooring, resilient flooring, carpeting, special flooring, floor treatment special coatings, painting, wall covering.

DIVISION 10—SPECIALTIES. Specialty items such as chalkboards and tackboards, compartments and cubicles, louvers and vents that are not connected with HVAC ductwork, wall and corner guards, access flooring, specialty modules, pest control, fireplaces, flagpoles, identifying devices, pedestrian control devices, lockers, protective covers, postal specialties, partitions, scales, storage shelving, wardrobe specialties, and the like.

DIVISION 11—EQUIPMENT. The electrical contractor may have need to refer to this section for equipment details requiring electrical connections, although the items themselves are indicated on the working drawings and possibly in Division 16 of the construction specifications. Types of equipment will include vacuum cleaning system (probably furnished by others and wired by the electrical contractor), bank and vault equipment, commercial equipment, checkroom equipment, darkroom equipment, ecclesiastical equipment, educational equipment, food service equipment, vending equipment, athletic equipment, industrial equipment, laboratory equipment, laundry equipment, library equipment, medical equipment, mortuary equipment, musical equipment, parking equipment, waste-handling equipment, loading dock equipment, detention equipment, residential equipment, theater and stage equipment, and registration equipment.

From the preceding list, it is easy to see that most of the items listed will require electrical connections. The equipment, however, will almost always be furnished by the general contractor, the owners, or others. The drawings and Division 16 should be checked to see who is responsible for the storage, moving, and setting in place.

Sometimes it is the responsibility of the person or firm supplying the equipment to make final connections. In these cases, the installer will usually hire the electrical contractor on the project to wire the item at an extra charge.

DIVISION 12—FURNISHINGS. Items covered by this division seldom, if ever, involve the electrical contractor. Some of the items covered include artwork, cabinets and storage, window treatment, fabrics, furniture, rugs and mats, seating, and other similar furnishing accessories.

DIVISION 13—SPECIAL CONSTRUCTION. The electrical contractor and others who are responsible for the electrical work should glance through this division on every project, just to make sure nothing is omitted on the electrical drawings that may require electrical connections. Air-supported structures, for example, will in most cases require electrical current to drive the air compressor; incinerators will require electrical connection as well as items covered by this division. All of the connections should be clearly shown on the electrical drawings and/or listed in the written electrical specifications, but it would be wise to double check yourself in this division.

DIVISION 14—CONVEYING SYSTEMS. As the name implies, this division covers conveying apparatus such as dumbwaiters, elevators, hoists and cranes, lifts, material-handling systems, turntables, moving stairs and walks, pneumatic tube systems, and powered scaffolding. Since most of these items will require electrical connections, the electrical contractor should ascertain whose responsibility the connections are.

DIVISION 15—MECHANICAL. In general, this division of the construction specifications covers all the work of the heating, ventilating, air conditioning, and plumbing contractors. Much of the heating and cooling equipment will require electrical connection and should be so specified in Division 16 or shown on the electrical drawings. Usually the mechanical contractor is responsible for installing all of his equipment, furnishing all motor control equipment (although there are exceptions), and so on. The electrical contractor usually is responsible for furnishing all materials and labor to install the main feeder to the pieces of equipment, as well as the final connection. The mechanical contractor is in most cases responsible for all control wiring to properly operate the equipment.

DIVISION 16—ELECTRICAL. All items under this division concern the electrical contractor. A good set of electrical specifications will clearly indicate exactly what work is the responsibility of the electrical contractor, and what work is not his responsibility. Unfortunately, this is not always the case; conflicts between the divi-

Figure 15.1. On an electrical installation of power wiring for this trailer park, who is responsible for removing the trees to install the underground cable? The electrical specifications should clearly state who.

sions exist as well as conflicts between the electrical drawings and the electrical specifications. Details of the most common conflicts as well as other explanations of the electrical specifications are given in the paragraphs to follow.

15.2 General Conditions

The General Conditions or Provisions section of Division 16 (electrical) consists of a selected group of regulations that apply to all subdivisions of the electrical work on the project. Depending on the size, type, and complexity of the project, these provisions may involve only a few concise paragraphs or could be comprised of several pages of detailed instructions.

In nearly all cases, the first few paragraphs of the General Conditions remind those persons using the specifications to refer to the architectural general and special conditions as they will be a

Figure 15.2. When the specifications recommend that the contractor examine the job site—especially on existing work—prior to bidding, the contractor should do so; the working conditions could greatly affect the labor. The vats in the floor of this tannery would greatly hinder the installation of lighting fixtures in the ceiling.

2. WORK INCLUDED (continued):

(D) The entire work provided for in this Section of the Specifications shall be constructed and finished in every part in a good, substantial and workmanlike manner according to the accompanying Drawings and Specifications to the full intent and meaning thereof. All work to be done in a finished manner. It is the intention of these Specifications particularly that the smaller details necessary for a workmanlike job and not usually included in the Specifications or indicated on the Drawings are to be included.

(E) Electrical Work for Heating, Ventilating, Air Conditioning and Plumbing:

(1) The Heating, Air Conditioning and Ventilating Contractor will furnish electrical equipment that is an integral part of the mechanical equipment, i.e., control panels on air conditioning equipment, thermostats on unit heaters, self-contained control equipment.

(2) The Heating, Air Conditioning and Ventilating Contractor shall install the temperature control system for the heating, air conditioning and ventilating system. The balance of the electrical work shall be performed by the Electrical Contractor. The Contractor shall refer to the Heating, Air Conditioning and Ventilating Section of these Specifications.

(3) The Electrical Contractor shall connect to all heating, air conditioning and ventilating and plumbing equipment.

(4) The Electrical Contractor shall furnish and install control wiring between the air conditioning condensing units and the air handling units.

(5) The Electrical Contractor shall furnish fused or unfused disconnect switches ahead of all equipment where required by the equipment supplier or the Owner. Combination starters may be used where applicable.

16010 Page 7

Figure 15.3. Sample page from an actual electrical specifications.

part of the electrical specifications. The work included follows, but again the details of this vary considerably from project to project. For example, some specifications merely give only a general description of the work, because more detailed data follow in subsequent pages. Other sets, however, give a more detailed description.

Following the "work included" section, an outline of the work not included should appear. These paragraphs tell the contractor what electrical equipment items are to be furnished by others, but installed by the electrical contractor; and what equipment is to be furnished *and* connected by others.

Information pertaining to codes and fees, tests, identification, and demonstration of a complete electrical system is standard in this portion of the specifications. If the electrical contractor is responsible for providing a temporary electrical service during construction, details of this service should also appear in the General Provisions of the electrical specifications.

15.3 Interpreting Electrical Specifications

In general, electrical specifications give the grade of materials to be used on the project and the manner in which the electrical system shall be installed. Most specification writers use an abbreviated language; although it is relatively difficult to understand at first, experience makes possible a proper interpretation with little difficulty. However, during the bidding stage of the project, the contractor, estimator, and others who will be involved in the project should make certain that everything is clear. If it is not, contact the architectural or engineering firm and clarify the problem prior to bidding the work, not after the work is in progress.

The following is an outline of the various sections normally included in Division 16 of the written construction specifications.

16010 GENERAL PROVISIONS. The General Provisions of the electrical specifications consists of a selected group of considerations and regulations that apply to all sections of this division. Other items included in this section have been discussed previously.

16100 BASIC MATERIALS AND METHODS. The contractor should look for clauses in this section that establish a means of identifying the type and quality of materials and equipment to be used in the project's electrical system. This section should further establish the accepted methods of installing the various materials.

16200 POWER GENERATION. Items covered in this section vary from job to job, but usually cover items of equipment used for emergency or standby power facilities; the type used to take over essential electrical service during a normal power failure.

16300 POWER TRANSMISSION. This section deals mainly with high voltage (over 600 volts) power transmission circuits and is included in projects constructed on government reservations and large industrial sites. For other projects, this type of work is usually handled by the local power company.

16400 SERVICE AND DISTRIBUTION. Power distribution facilities (under 600 volts) are covered in this section of the electrical specifications by descriptive paragraphs or clauses covering selected related equipment items.

16500 LIGHTING. This section of the specifications normally covers general conditions relating specifically to the selected lighting equipment to insure that all lighting equipment is furnished and installed exactly as the designer selected and specified. Further clauses establish the quality and type of all lighting fixtures, accessories, lamps, and so forth. Methods of installation are also included in most sets of electrical specifications.

16600 SPECIAL SYSTEMS. Such items as lighting protection equipment, special emergency light and power systems, storage batteries, battery charging equipment, and perhaps cathodic protection are examples of items that could be covered in this portion of the specifications.

16700 COMMUNICATIONS. Equipment items that are interconnected to permit audio or visual contact between two or more stations, or to monitor activity and operations at remote points are basically the items found in this portion of the specifications. Most clauses are designed around a particular manufacturer's equipment stating what items will be furnished and what is expected of the system once it is installed and in operation.

16850 HEATING AND COOLING. Because of working agreements between labor unions, most heating and cooling equipment, with few exceptions, are installed by workmen other than electricians and the requirements are usually covered in Division 15 of the specifications. However, in certain cases, the electrical contractor will be responsible for installing certain pieces of the equipment, especially on residential and apartment projects. This section of the electrical specifications covers all necessary details pertaining to the equipment and the installation when such a condition exists.

16900 CONTROLS AND INSTRUMENTATION. As the name implies, this section of the specifications covers all types of controls and instrumentation used on a given project. Some of the items covered include: recording and indicating devices, motor control centers, lighting control equipment, electrical interlocking devices, and applications, control of electric heating and cooling, limit switches, and numerous other such devices and systems.

Other divisions of the construction specifications may involve a certain amount of control work, especially Division 15. The responsibility for such work should be clearly defined before bidding the project.

15.4 Degree of Compliance with Specifications

Most architects and consulting engineers will specify the type and design of certain items of electrical equipment (panelboards, motor control centers, lighting fixtures, and the similar items) by listing a given manufacturer's name and catalog number of the item. However, when items are specified in this manner, usually this particular make and model number, *or its approved equal*, will be accepted.

Some architects, engineers, and owners will be very reasonable when it comes to approving other than specified makes of items; others may require such close compliance that no other manufacturer's item will be accepted as an "equal." This is especially true for lighting fixtures, panelboards, and certain types of special equipment. Many of these problems occur when the architect/engineer allows a particular manufacturer to write a portion of the specifications for the firm. In this case, the manufacturer is certain to write an iron-clad specification which only his (the manufacturer's) equipment will meet.

The architect/engineer may specify a particular wiring method, or may specify wiring methods in general, and then require the contractor to use the most expensive of the group.

15.5 Common Conflicts and How to Handle Them

The contractor and his workmen should always be on the alert for conflicts between working drawings and the written specifications, or between Division 16 and other divisions of the construction documents. Such conflicts occur particularly when:

1. Architects or engineers use standard or prototype specifications and attempt to apply them to specific working drawings without any modification.
2. Previously prepared standard drawings are to be changed or amended by reference in the specifications only; the drawings themselves are not changed.
3. Items are duplicated in both the drawings and specifications and then an item is amended in one or the other and overlooked on the other contract document.

An example of item No. 3 would be a power-riser diagram shown on a sheet of the working drawings, indicating diagrammatic locations of all panelboards and related service equipment to be used on the project. The written specifications list all of the panelboards including their contents (fuses, circuit breakers, etc.) in a panelboard schedule. If another panel must be added for the project at a later date, prior to the job going out for bids, it will most often be added to the power-riser diagram on the drawings, but such a change is often overlooked in the written specifications, especially if it is a last-minute change. For this reason, it is best not to duplicate items in the specifications and on the drawing. Rather, the proper place should be determined for the information to be listed, and then indicate it in one or the other of the construction documents, not in both.

In such instances, it is the responsibility of the person in charge of the project to ascertain which takes precedent over the other;

that is, the drawings or the specifications. When such a condition exists, the matter must be cleared up, preferably before the work is installed, in order to avoid added cost to either the owner, architect/engineer, or the contractor.

15.6 Substitution for Specified Items or Methods

Electrical suppliers often offer quotations on substitute items at a considerably lower price than the named item, implying that the substitute item is equal to the named item. It is a responsibility of an estimator to determine whether a substitute item would be accepted by the architect or engineer, particularly if it is priced lower than the make named in the specifications. This is normally handled by the submittal of shop drawings during the construction sequence. However, if the contractor is to use the reduced prices as a basis for his bid the approvals should be made prior to the bid date.

When the contractor has based his bid on an unauthorized substitution, electrical contractors many times find themselves faced with furnishing items required by the architect or engineer at a higher cost than was used in making up the estimate. Some electrical contractors will purposefully base their estimates on lower-priced items of material and equipment not equal to those specified, and then attempt to obtain approval of the substitutions. While in some instances they may be able to do so, in more cases than not they are forced to furnish the items as specified. Such practices are to be condemned, as such actions spoil the project for themselves as well as for the other bidders; this practice further destroys the confidence of the customers, architects, and engineers in the electrical contracting industry.

Sometimes the architect or engineer is willing to accept a substitute item or wiring method when certain economies can be effected and the quality or efficiency of the electrical installations will not be impaired; however, they usually require a credit of the difference in the costs of the items or the savings brought about through the change in the wiring method.

A typical paragraph from an electrical specification (concerning the substitution of materials) may read as follows:

The naming of a certain brand or make or manufacturer in the Specifications is to establish a quality standard for the article desired. The

Contractor is not restricted to the use of the specific brand of the manufacturer named unless so indicated in the Specifications. However, where a substitution is requested a substitution will be permitted only with the written approval of the Engineer. No substitute material or equipment shall be ordered, fabricated, shipped, or processed in any manner prior to the approval of the Architect–Engineer. The Contractor shall assume all responsibility for additional expenses as required in any way to meet changes from the original material or equipment specified. If notice of substitution is not furnished to the Architect-Engineer within ten (10) days after the Contract is awarded, then equipment and materials named in the Specifications are to be used.

15.7 Compliance with Codes and Standards

The electrical contractor should carefully study the electrical specifications to see who is responsible for obtaining and paying for all permits, inspection fees, and installation fees required to complete the electrical work as called for in the drawings and specifications. The instructions are usually included under the General Provisions of Division 16.

Where the project will be serviced by local utility companies, the power company should be consulted, prior to bidding, ascertain that their requirements are understood and that the prices for accomplishing the requirements are included in the bid.

All materials and workmanship must comply with all applicable codes, drawings, specifications, local ordinances, industry standards, utility company, and fire insurance carriers requirements.

In case of differences between the building codes, plans, specifications, federal laws, state laws, local ordianaces, industry standards, utility company regulations, fire insurance carrier's requirements and the contract documents, the most stringent usually will govern. The contractor should promptly notify the architect–engineer in writing of any such difference.

When the contractor performs any work that does not comply with the requirements of the applicable building codes, state laws, local ordinances, industry standards, fire insurance carrier's requirements, and utility company regulations, he will almost always be required to bear the cost arising in correcting any such deficiency.

Applicable codes and all standards will include state laws, local ordinances, utility company regulations and the applicable requirements of the following nationally accepted codes and standards:

a. ***Building Codes:***

1. National Building Code
2. Local Building Code
3. National Electrical Code
4. State Electrical Code

b. ***Industry Standards, Codes and Specifications:***

1. ASTM—American Society for Testing and Materials
2. EEI—Edison Electric Institute
3. IEEE—Institute of Electrical and Electronic Engineers
4. NBS—National Bureau of Standards (National Electrical Safety Code)
5. NEC—National Electrical Code (NFPA No. 70-1978)
6. NFPA—National Fire Protection Association
7. UL—Underwriters' Laboratories, Inc.
8. USASI—United States of America Standards Institute

c. ***Rules and Regulations of the Local Electric Utilities Where Applicable.***

16

Fundamental Estimating Techniques

Those persons engaged in the highly competitive business of electrical contracting know that the only way to continue profitable operations is to practice extremely efficient management techniques; use only workmen specifically trained or experienced and adapted to the type of electrical work performed; and develop an accurate and relative fast system of estimating electrical work. This chapter covers some essentials necessary for proper estimating of electrical work.

16.1 Estimating Facilities

The electrical contractor's office, whether large or small, is required to take up at least the following functions:

1. Business administration.
2. Supervision and job management planning.
3. Design (to some extent).
4. Estimating.

A typical office may occupy a single room where all activities are carried out, or if could occupy an entire floor or even a whole building. Factors that dictate the size of the contractor's office include volume of business, method in which the contractor operates his business, and the availability of suitable space.

Estimating facilities will also vary, but in general, a separate space should be provided that is out of the way of the usual office routine. It should be illuminated with approximately 100–150 footcandles of well-diffused light; painted a pleasing color (preferably pastel shades); be provided with an adequate heating, cooling, and ventilating system; and be of sufficient size to accommodate all the necessary estimating tools and materials.

The furnishings within the drafting room should consist of a drawing board, utility or throw-off table or desk, comfortable stools or chairs, filing cabinets and plan files. Bookshelves for manufacturer's catalogs, estimating manuals, and pricing data should be placed near the desk within easy reach of the estimator.

Each estimator should be provided with a reliable electronic calculator. The use of such calculators cuts down on human error, helping to eliminate costly mistakes. Savings in estimators' time will pay for the calculator many times over.

The estimator should also be equipped with drafting instruments and supplies, a rotameter, metal tape graduated to ⅛- and ¼-inch scales, and various colored pencils; a tabulator for counting outlets and other electrical components should also be provided. Circuit and feeder lines on scaled drawings are usually measured with the rotameter, which indicates the actual footage of the run when graduated to the proper scale. The colored pencils are helpful in checking or indicating the outlets, circuits, and other electrical equipment as they are taken off and accounted for on takeoff forms or worksheets.

For significantly increased speed and accuracy in electrical estimating, there are several highly sophisticated estimating calculators on the market. These calculators are capable of measuring linear and area dimensions and of making quantity counts from drawings drawn to any scale. On one type, the estimator, by manipulating two probes, can automatically accumulate and display the lineal and area measurements and quantity counts, which can then be used for immediate electronic extension by cost and labor factors.

The flexibility of these calculators makes them especially useful to electrical estimators and cost engineers. The cost of these instruments normally starts out about $800 and rises to a price of over $2000. Some are even equipped with colored ink pens to automatically mark the circuit runs or outlets as they are being taken off the drawing.

Two drafting facilities are shown in Figs. 16.1 and 16.2; the first would be used by a one-man firm, the second by a larger firm

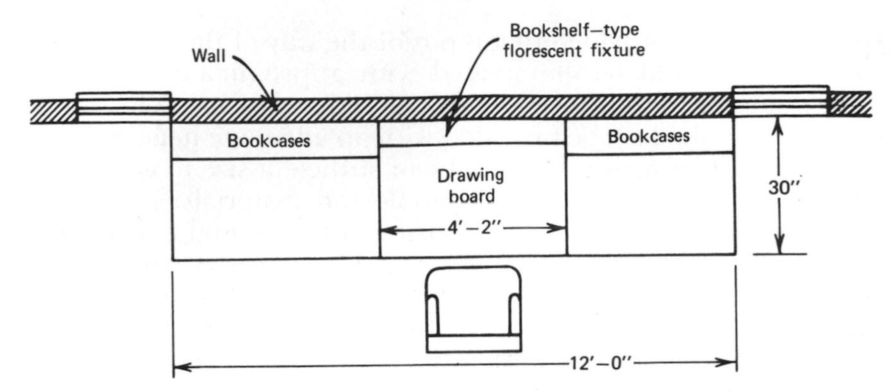

Figure 16.1. Estimating facility for small one-man contracting firm.

hiring several estimators, designers, and draftsmen. A study of these illustrations should help the reader decide upon a suitable layout for his own estimating facilities.

Take, for example, the floor plan in Figure 16.2. This estimating room is large enough to accommodate four estimators by using "U" shaped counters along one entire wall, this gives three of the estimators an L-shaped work area. Drawing boards are mounted on one leg of the L-shaped counter tops while the remaining space is used for reference materials or throw-off space.

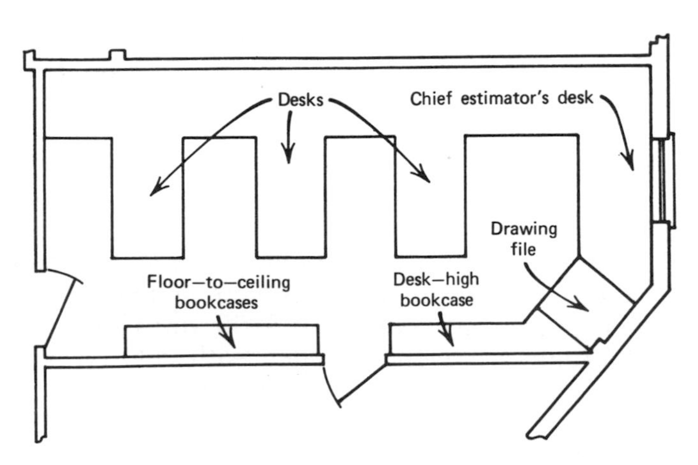

Figure 16.2. Estimating facilities for four estimators when space is limited.

Duplex receptacles are located above the countertops along the wall to provide electricity for electrically operated devices such as electric erasers, and calculating machines.

The wall opposite the counters contains bookshelves for storage of manufacturers' catalogs and equipment data. At the end of each counter (containing the drawing boards) is a built-in bookcase the same width and height as the counter itself. These bookcases contain reference books most frequently used by the estimator at each space: estimating manuals, design data, and so on.

Notice that the end area next to the window is somewhat larger than the three remaining areas. This area was planned for the chief estimator, who handles more drawings and must have more work space to organize the estimating procedures. The plan file, to his immediate right, contains drawings of projects under construction. He therefore has these drawings at hand when questions are phoned in by the workmen on the job.

To the immediate left of the drawing board is a large throw-off space with a plan file. The walls above the counter contain charts in large print of the most-used estimating and design data for quick reference. Two drawer legal-size filing cabinets installed under the countertops are used for storage of job files and other reference material.

The chief estimator is also provided with a telephone, tape recorder for dictation, an electronic calculator, and an assortment of other drafting and estimating tools and materials.

16.2 Estimating Manuals and Forms

Electrical estimates can be prepared for all types and sizes of electrical construction with remarkable accuracy, provided detailed and accurate material lists are prepared and proven labor units are applied. Of course, variable job factors must also be taken into consideration and good job management is required to complete the project at a profit.

The first step in arriving at the most accurate total estimated labor for a given job is to take off and list the various material items on pricing sheets, segregated in accordance with the installation and building construction conditions.

The second step is to apply the labor unit specifically related to that particular installation condition for the size and type of material involved. The correct labor unit for a given item may have been derived by the contractor's own experiences. However, most electrical contractors use labor unit manuals that have tested and proven accurate over a number of years. One of the earliest manuals of this type was prepared by the National Electrical Contractors Association for use by its members. In recent years, numerous other labor pricing manuals have appeared on the market and can be leased or purchased by any contractor.

It is recommended that electrical contractors examine all of the manuals that he can locate, and then decide which one best suits his own needs. One manual that has met with favorable response is distributed by Estimatic Corporation, Denver, Colorado 80223. Their system, known as Datamatic, was developed to fill the need for a simple, vast and reasonably accurate system for estimating electrical work.

This system utilizes data processing to keep material and labor values current and, at the option of the user, can produce through the computer in a matter of minutes, material and labor values which have been factored, extended, and totaled from a simplified take-off procedure. The simplified take-off procedure was developed by combining items of material that are commonly used together and assigning material and labor values to each group. This procedure greatly simplifies the take-off processes and provides an accurate basis for the estimate.

Material prices and labor hours are listed on charts for each estimating unit and these are kept up to date as the market changes by mailing out revised sheets to their subscribers.

Since this system utilizes the principal of averages and probability with an estimate with acceptable overall accuracy, it is a simple approach to estimating normal electrical work with fast and accurate results easily obtained. However, the system does have some limitations which must be respected in order to achieve the most satisfactory results. One of the most important limitations to consider is its use on specialized work; its greatest accuracy is realized under normal electrical work:

1. Complete electrical installations consisting of service entrance, distribution system, branch circuit work, motors, lighting fixtures and devices. Highly specialized work which does not include most of the preceding work items is likely to be less accurate.
2. Labor units are based on average installation conditions; not the easiest, nor the most difficult. Therefore, jobs that would be considered exceptionally easy or difficult must receive appropriate labor adjustments.

Bear in mind that this is not the only system on the market and others may suit your needs better. Check for advertisements in electrical trade journals for names and addresses of other firms issuing manuals of labor units; then, after examining them all, decide which method is best for you.

Many calculations and records must be made during the process of estimating electrical construction. Such calculations are best performed when a systematic pattern is followed, using appropriate forms. In general, all forms have spaces for the name of the project, the date, the names of the estimator, and other standard data. Forms discussed here include the following:

1. Pricing sheet.
2. Material and price sheet.
3. Estimate sheet.
4. Bid and estimate summary.
5. Bid summary sheet.
6. Small take-off and listing sheet.
7. Large take-off and listing sheet.
8. Branch circuit schedule.
9. Feeder schedule.
10. Conduit and wire summary.
11. House wiring summary.
12. House wiring recap.

All of these forms are available from the Minnesota Electrical Association, 3100 Humboldt Ave. South, Minneapolis, Minnesota 55408. A free catalog will be mailed to any electrical contractor on request.

PRICING SHEET. Pricing sheets (Figure 16.3) provide for material and labor listings as well as material prices and labor units. Separate pricing sheets should be used for each floor or section of the project and summarized on the bid summary sheet shown in Figure 16.7.

MATERIAL AND PRICE SHEET. Figure 16.4 shows another type of summary sheet that can save the electrical contractor much valuable time in listing each item. This form provides spaces for all commonly used materials, quantities consumed on the job, and the pricing of these items. This form is particularly adaptable to

PRICING SHEET

JOB ______________________ ESTIMATE NO. ________

WORK ______________________ SHEET NO. ____ OF ____ SHEETS

ESTIMATED BY ________ PRICED BY ________ EXTENDED BY ________ CHECKED BY ________ DATE ________

	MATERIAL	MATERIAL					LABOR			
		QUANTITY	LIST PRICE	PER	DISC.	EXTENSION	UNIT	PER	EXTENSION	
1										1
2										2
3										3
4										4
5										5
6										6
7										7
8										8
9										9
10										10
11										11
12										12
13										13

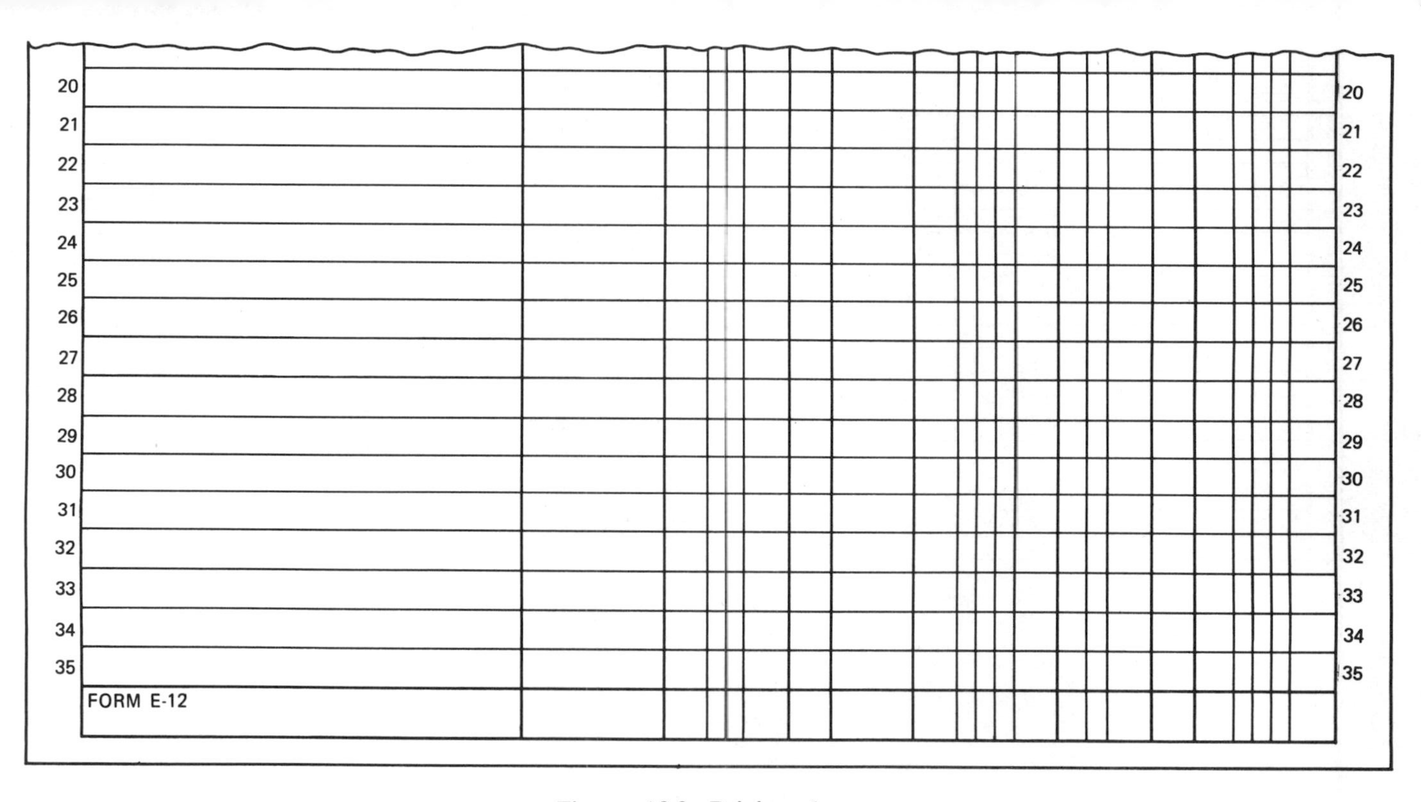

Figure 16.3. Pricing sheet.

MATERIAL & PRICE SHEET

JOB ______________________ DATE _____________

CONDUITS AND FITTINGS									
KIND	SIZE			SIZE			SIZE		
	QUAN	EACH	TOTAL	QUAN	EACH	TOTAL	QUAN	EACH	TOTAL
Rigid Conduit									
Conduit Ells									
F.E.B.									
L.A.Y.									
LB – L – R									
A – B									
Blank Cover									
Bushings									
Lock Nuts									
Steel Tube									
S.T. Couplings									
S.T. Connectors									
E.M.T. Straps									

OUTLET BOXES–COVERS				SWITCH CABINETS–DEVICES				100 Amp. Cart.			
KIND	QUAN	EACH	TOTAL	KIND	QUAN	EACH	TOTAL	200 Amp. Cart.			
Steel B. Box				30 A Pole Plug							
Steel Bracket				30 A Pole Cart				Bell Wire			
Steel Standard				60 A Pole Plug				Ins. Staples			
Steel Short				60 A Pole Cart				Transformer			
Handy Box				100 A Pole Cart							
Handy Covers				200 A Pole Cart				Rubber Cord			
4-11/16 Sq. $1\frac{1}{2}$								Tape			
4-11/16 Sq. $2\frac{1}{8}$				Range Cab. 4 Cir				BX Staples			
4-11/16 Rings								Cable Straps			
4" Oct. $1\frac{1}{2}$				Breaker Panel							
4" Oct. $2\frac{1}{8}$				Breaker SP				Exp. Bolts			
4" Oct. Rings				Breaker DP				Drive–in Plugs			
4" Square $1\frac{1}{2}$				SP Flush Sw.				Toggle Bolts			
4" Square $2\frac{1}{8}$				SP Surf. Sw.							
4" Sq. Rings				3w Flush Sw.				Connectors			
4" Sq. Covers				4w Flush Sw.							
				Duplex Recpt.							
4" Porc. Pull				We Prf. Recpt.							
								Total			

Conduit Couplings								
Totals								

WIRE									
KIND	NO.			NO.			NO.		
	QUAN	EACH	TOTAL	QUAN	EACH	TOTAL	QUAN	EACH	TOTAL
Rubber Cov.									
Weatherproof									
Non–Met. 2 Wire									
Non–Met. 3 Wire									
Non–Met. w/Grd.									
BX 2 Wire									
BX 3 Wire									
S.E.C. 2 Wire									
S.E.C. 3 Wire									
Lead 2 Wire									
T.W.									
Totals									

Bar Hangers			
Fixture Studs			
Total			

GROUNDING AND HARDWARE			
KIND	QUAN	EACH	TOTAL
Grd. Rods			
Grd. Clamps			
Grd. Straps			
Insulators–Scr			
2 Wire Racks			
3 Wire Racks			
Lag Screws			
Mach. Bolts			
Dead Ends			
Total			

Pilot & Switch			
Sw. Plate–			
Sw. Plate–			
Sw. Plate–			
Recp. Plate–			
Recp. Plate–			
Cabinet–			
Cabinet–			
Total			

MISCELLANEOUS AND FIXTURES			
KIND	QUAN	EACH	TOTAL
Fustats, 1–14			
Fustats, 15–30			
Adaptors			
30 Amp. Cart.			
60 Amp. Cart.			

RECAP	
ITEM	AMOUNT
Conduits & Fittings–	
Conduits & Fittings–	
Conduits & Fittings–	
Wire, No.–	
Wire, No.–	
Wire, No.–	
Boxes & Covers	
Grndg. & Hdwe.	
Sw. Cab's & Dev.	
Misc. & Fixtures	
Total Material	
Labor	
Total	
Overhead	
Profit	
Total	
Sales Tax	
Total	
Sell For–	

Figure 16.4. Material and price sheet.

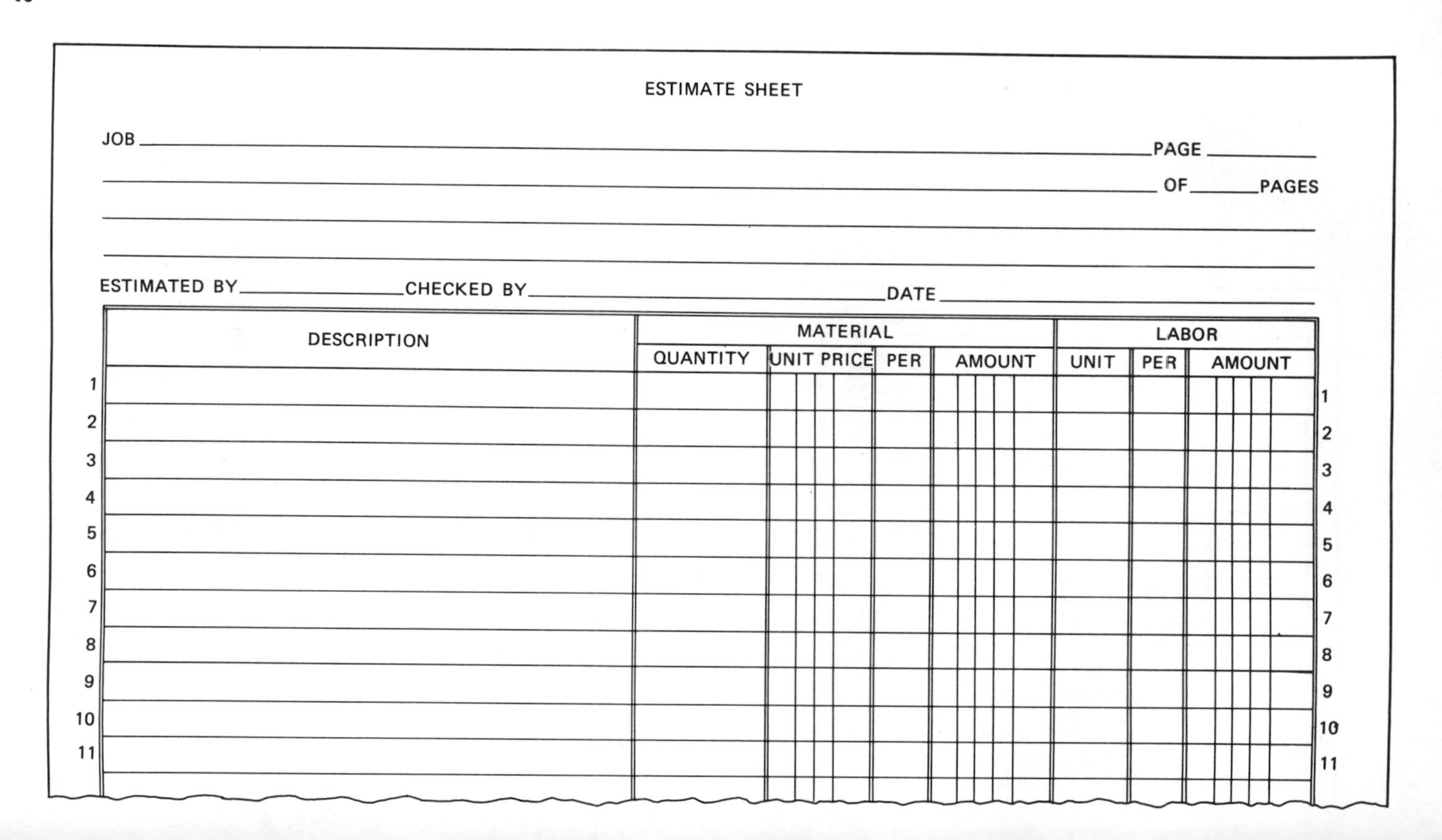

ESTIMATE SHEET

JOB ______________________ PAGE ______

______________________ OF ______ PAGES

ESTIMATED BY ______ CHECKED BY ______ DATE ______

	DESCRIPTION	MATERIAL				LABOR			
		QUANTITY	UNIT PRICE	PER	AMOUNT	UNIT	PER	AMOUNT	
1									1
2									2
3									3
4									4
5									5
6									6
7									7
8									8
9									9
10									10
11									11

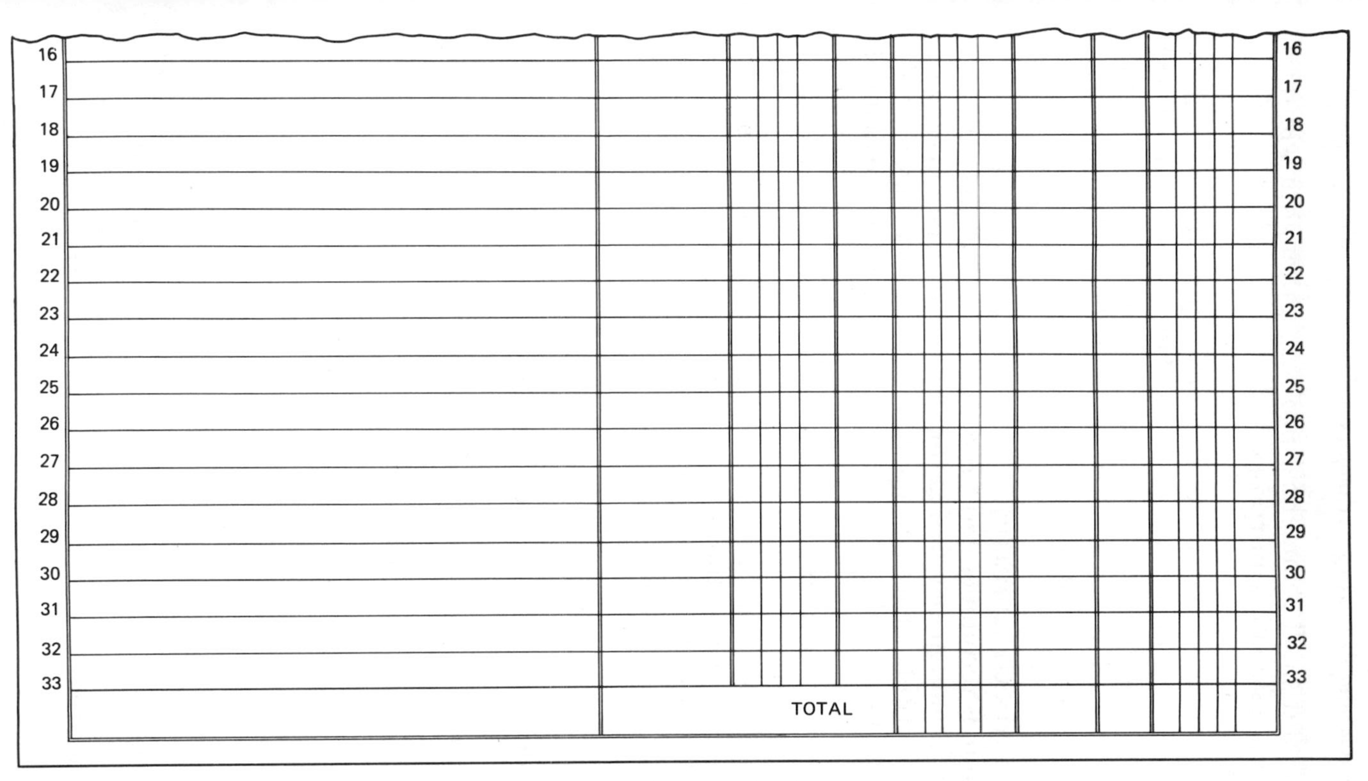

Figure 16.5. Estimate sheet.

residential and small commercial projects, although it will serve most general estimating purposes.

ESTIMATE SHEET. The estimate sheet in Figure 16.5 is used for precise estimating in making electrical bids. It is arranged with description, material and labor columns which allow orderly recording of costs. The use of separate estimate sheets is suggested for each floor or section of the job.

BID AND ESTIMATE SUMMARY. This complete summary sheet (Figure 16.6) is recommended for electrical contractors who want detailed information included in the sheet format, eliminating the necessity of writing each item separately. The front (Figure 16.6a) is used to summarize all costs in making a bid; that is, the first half provides space to list each separate estimate sheet and put in the total of material and labor. When columns for material and labor are totaled, the next calculation will be nonproductive labor and job expense. These calculations are made easy by using the columns located at the bottom of the sheet in Figure 16.6a, taking into consideration all costs and profit to arrive at a total price. The reverse side (Figure 16.6b) of the bid and estimate summary sheet has spaces for direct job expense, and other pertinent information.

BID SUMMARY SHEET. The bid summary sheet (Figure 16.7) is basically the same as the bid and estimate summary sheet except the bid summary sheet has a complete checklist on the reverse side to help the electrical contractor avoid some of the common errors made in estimating electrical systems.

SMALL TAKE-OFF AND LISTING SHEET. As the length of circuit runs, conduit, wire, outlet boxes, and so on are measured and counted from the plans, they must be listed first to obtain total quantities of each item or type of material before being transferred to regular estimate or pricing sheets. The form in Figure 16.8 provides an orderly means of performing this operation. In the squares at the top of the columns (between the double-ruled lines) the type and size of material is listed. After the required materials are so listed, totals can be transferred to the estimate sheet (Figure 16.5).

LARGE TAKE-OFF AND LISTING SHEET. This form (Figure 16.9) is used for the listing of various materials as they are measured or counted from plans, using one cross line for each circuit or each room as desired. The various items of material, conduit and wire (by sizes) outlet boxes, wiring devices, and so on are to be entered in the top row of slant lines and the quantities of each listed in the vertical column under each item. Listing space is provided for 33 items and 30 rooms or sections.

BRANCH CIRCUIT SCHEDULE. A self-explanatory form used for listing and summarizing outlets, switches, wall and ceiling pipe entrances, and conduit and wire lengths and measurements is shown in Figure 16.10. This contains space for job location and all other pertinent information.

FEEDER SCHEDULE. The form in Figure 16.11 has necessary columns for feeder numbers and from beginning of feeder conduit to end connection. Also has columns for conduit size, length, loads, terminals, and bends; wire size length; and number of wires to each size conduit.

CONDUIT AND WIRE SUMMARY. An essential form used in listing sizes, lengths, prices, types, and any other information pertinent to the estimation of required conduit and wire materials is shown in Figure 16.12. This form included space for necessary job numbers, dates, and so forth.

HOUSE WIRING SUMMARY. A simple, complete outline on one sheet for a full summary of equipment and wiring for a house wiring job is shown in Figure 16.13. The full information at a glance, containing 57 items and easily fileable, is included on the form. It can be used to tell the contractor at a glance just what was included in a residential bid, and it can then be used for a checklist during the construction. Full-length columns provide for both material and labor.

HOUSE WIRING RECAP: The form in Figure 16.14 provides a simple checklist for estimating residential construction and can also be used as a checklist after installations have been made.

16.3 Automatic Estimating System

Automatic estimating systems have been used in the larger contracting firms for a number of years, but now the lower cost of solid-state circuitry has made such systems within the means of

BID AND ESTIMATE SUMMARY

PROJECT NAME ______ ESTIMATED BY ______ ESTIMATE NO. ______

LOCATION ______ CHECKED BY ______ SHEET NO. ___ OF ___ SHEETS

______ APPROVED BY ______ DATE ______

SCHEDULE I – SUMMARY OF ESTIMATE SHEETS

SHEET NUMBER	SECTION	MATERIAL	LABOR IN HOURS OR DOLLARS
	TOTALS		

SCHEDULE II – LABOR COSTS BY DOLLARS OR MAN–HOURS

(USE LIGHT AREAS FOR FIGURES)

Total Labor from Schedule I		
Labor Job Factor using Percent of Total Labor in Schedule I:		
Weather @—%(0.20%)		
Size of Job @—%(0.30%)		
Coordination @—%(0.10%)		
Complexity @—%(0.15%)		
Labor Efficiency Factor @—%(± ? %)		
Stand-By (Lump Sum)		
Total Job Factor		
Non-Productive Labor:		
Supervision		
Timekeeper-Stockman-Job Clerk		
Ordering and/or Handling Material		
Lost Time and/or Traveling Time		
Testing if Required or Desirable		
Other –		
Total Non-Productive Labor		
Total Labor in Man-Hours or Dollars		
(If labor has been figured in man-hours, multiply total above by average labor rate per hour $_____)		

Figure 16.6a. Bid and estimate summary (front side).

SCHEDULE III – RECAP		(USE LIGHT AREAS FOR FIGURES)	
Total Labor Cost from Schedule II			
Employers Labor Expense Based on Labor Cost (above)			
Social Security	@—%		
Unemployment Insurance	@—%		
Workman's Compensation Insurance	@—%		
Public Liability & Property Damage Insurance	@—%		
Pension Fund	@—%		
Local Health and Welfare Insurance	@—%		
Other –	@—%		
Total Employers Labor Expense			
Cost of Material from Schedule I			
Direct Job Expense:			
Bid Bond			
Insurance (special)			
Inspection & Permit Fees			
License			
Extra Engineering & Drawings			
Job Office & Storage Shed			
Watchman			
Freight, Cartage & Storage			

On-Job Truck Expense		
Telephone		
Power-Light-Heat		
Tool and Equipment Expense		
Travel Expense – ____ days or miles @ ____		
Room and Board – ____ days @ ____		
Other –		
Total Direct Expense		
Total Prime Cost		
Overhead Based on Prime Cost @ ____ %		
Total Net Cost		
Profit Based on Net Cost @ ____ %		
Total This Estimate		
Plus Tax, Bond, Etc.:		
Sales Tax		
Payment and Performance Bond		
Other –		
Total		
Total Estimated Price		
Amount of Bid		

Figure 16.6b. Bid and estimate summary (back side).

BID SUMMARY SHEET

JOB ______________________________ SHEET NO. ____ OF ____ SHEETS

ESTIMATED BY ____________ CHECKED BY ____________ DATE ____________

SHEET NO.	DIVISION	MATERIAL--DOLLARS	LABOR--HOURS

NON-PRODUCTIVE LABOR	HOURS	Miscellaneous Material and Labor	
Handling Material		Non-Productive Labor	(A)
Superintendent		TOTALS – MATERIAL (C) & LABOR (D)	
Traveling Time and Lost Time		(D) ___ Hours Labor @ ___	
Job Clerk		___ Hours Labor @ ___	
TOTAL (A)		___ Hours Labor @ ___	
JOB EXPENSE	DOLLARS	Taxes: Soc. Sec. ___ Unemp. ___	
Tools, Scaffolds		Workmen's Compensation Insurance	
Pro Rata Charges		LABOR COST GROSS TOTAL	
Insurance, Public Liability, Etc.		Job Expense (B)	
Cutting, Patching, Painting		Material Cost (C)	
Watchman		TOTAL PRIME COST	
Telephone		___ % Overhead	
Drawings		TOTAL NET COST	
Inspection and Permit Fees		___ % Profit	
License		Selling Price Without Bond	
Storage		Bond	
Freight, Express and Cartage		Selling Price With Bond	
Transportation		PRICE QUOTED	
Board ___ Men ___ Weeks At ___			
TOTAL (B)			

To Avoid Errors, Check List on Reverse Side　　Let's Upgrade Our Electrical Industry

Figure 16.7a. Bid summary sheet (front side).

CHECKING LIST FOR BID SUMMARY AND ESTIMATE SHEETS

To avoid or catch some of the common errors or omissions in making up estimates and bids, check through this list. The short time required may save you money and embarassment. Make sure that all appropriate items are included in your bid properly figured.

CONDUIT, ETC. Sizes	⅜"	½"	¾"	1"	1¼"	1½"	2"	2½"	3"	3½"	4"	
Rigid Heavy												
Thin Wall Steel Tube												
Greenfield												
Conduit Ells												
Couplings												
Connectors												Straps
Lock Nuts—Bushings												Hangers

CONDUIT FITTINGS .... Regular .... Explosion Proof .... Dust Tight .... Water Tight .... Adapters ....

RACEWAYS .... Metal Mold .... Wire Mold .... Window Strip .... Wire Ducts .... Trough .... Under Floor .... Fittings ....

WIRE Single Cond.	16	14	12	10	8	6	4	2	0	2/0	3/0	4/0	250M	300	400	500
Code R. C.																
R W Moisture Resistant																
R H Heat Resistant																
R P Performance Type																
Weatherproof																

Racks .... Insulators .... Connectors .... Lags .... Bolts ....

CABLES Sizes	12	10	8	6	4	2	0	2/0	3/0	4/0	250M	Larger	1	2	3	4	Conductor
Lead Cover R. C.																	
Underground, Parkway, Etc.																	
Service Entrance																	

Potheads .... Service Heads .... Junction Boxes .... Splices ....

Non-Metallic (Inside) . . . | Cable Hangers ...... Straps ...... Sill Plates ......

ABC - BX (Inside) . . . . | Straight Connectors ...... Angle Connectors ......

CORDS—Lamp ...... Heater ...... Light Rubber ...... Heavy Rubber ...... Sockets ...... Guards ...... Attachments.

PANELS—Distribution ...... Light ...... Heat ...... Power ...... Signal ...... Cabinets ...... Pull Box ...... Cutouts ...... Meter Board ......

SAFETY SWITCHES—Meter ...... Entrance ...... Motor ...... Oil Switch ...... Relays ...... Time Switch ......

FUSES—Plug ...... Non-Tamperable ...... Nec. Cartridge ...... One Time ...... Renewable ...... Extra Sets ...... Links ......

GROUND—Rods ...... Pipe ...... Clamps ...... Fittings ...... Wire ...... Guards ...... Hardware ...... Exp. Bolts ...... Toggle Bolts ......

OUTLET BOXES—All Types and Sizes ...... Plaster Rings ...... Covers ...... Extensions ...... Fixture Studs ...... Hangers ......

WIRING DEVICES—1 ...... 3 ...... 4-Way Switches ...... Dupl. Recepts. ...... Plates ...... Spec. Finish ...... Special Devices ......

SIGNAL—Chime ...... Bells ...... Buzzer ...... Push Buttons ...... Board ...... Transformers ...... Wire ...... Batteries ...... Rectifier ......

CLOCKS—TELEPHONE—FIRE—BURGLAR—THERMOSTATS—Conduits ...... Wire ...... Outlets ...... Equipment ......

PUBLIC ADDRESS—RADIO—Equipment ...... Speakers ...... Conduits ...... Outlets ...... Shielded Wire ...... Antenna ......

MOTORS—Starters ...... Controllers ...... Push Button Stations ...... FANS—Vent. ...... Exhaust ...... Attic ...... Ducts ...... Louvres ......

FIXTURES—Typical ...... Special ...... Recessed ...... Show Wind ...... Stage ...... Outdoor ...... Flood ...... Guards ......

LAMPS—Mazda ...... Silv. Bowl ...... Flourescent ...... Lumiline ...... Pilot ...... Neon ...... Special ......

AUXILIARIES—Power Factor Correction ...... Capacitors ...... Starters ...... PORCELAIN ......

MISCELLANEOUS—Tape ...... Solder ...... Screws ...... Nails ...... | STRUCTURAL IRON, Etc. ......

HAVE YOU DOUBLE CHECKED PLANS FOR ALL DETAILS?

Service Entrances, Location	OUTLETS	All Floors, Basement, Attic.
Switches, Panels, Cabinets	Ceiling	Outside or Underground Wiring.
Feeder Runs, Sub Feeders	Wall Lights	Connecting All Motors, Etc.
Branch Circuits, Wiring System	Switch	Connecting All Equipment to be supplied by others.
Specified Materials & Methods	Floor	Addenda, Changes, Corrections.
Material Supplied by Owner	Receptacles	Assembling and Hanging Fixtures.
Labor Rates and Conditions	Special	

ARE ALL EXTENSIONS FIGURED PROPERLY? Per Each, Doz., 100 (C), Gross, 1000 (M), Feet, Lbs. Etc.
READ BID FORMS CAREFULLY - - - FILL IN AND SIGN PROPERLY
DON'T FORGET CERTIFIED CHECK OR BIDDERS BOND, IF REQUIRED

Figure 16.7b. Bid summary sheet (back side).

SMALL TAKE–OFF & LISTING SHEET

JOB __________ PAGE ______

__________ OF ______ PAGES

ESTIMATED BY ______ CHECKED BY ______ DATE ______

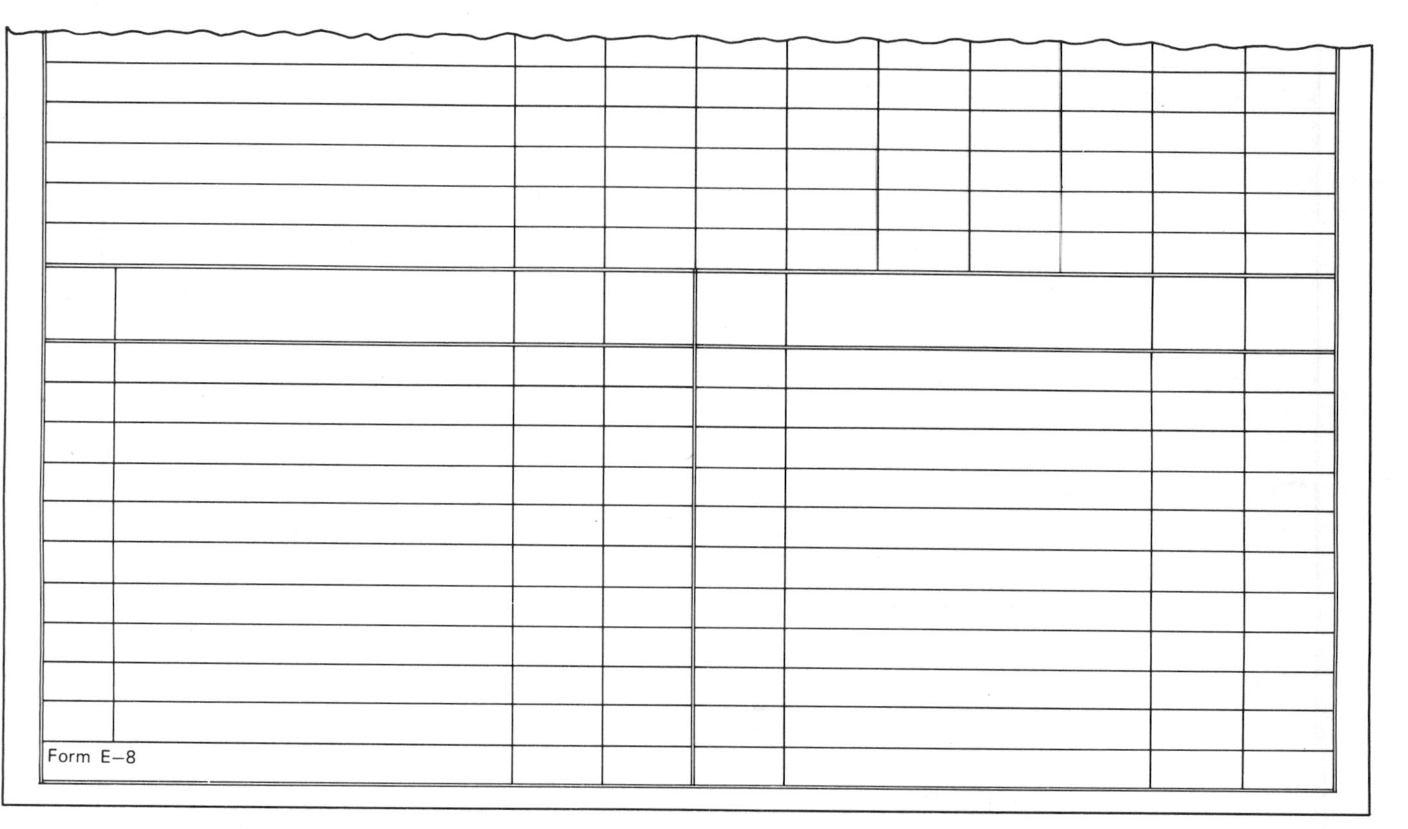

Figure 16.8. Small take-off and listing sheet.

LARGE TAKE–OFF & LISTING SHEET

JOB ____________________ ESTIMATED BY _____ PAGE ___ OF ___ PAGES

____________________ CHECKED BY ______ DATE __________

TOTALS

Figure 16.9. Large take-off and listing sheet.

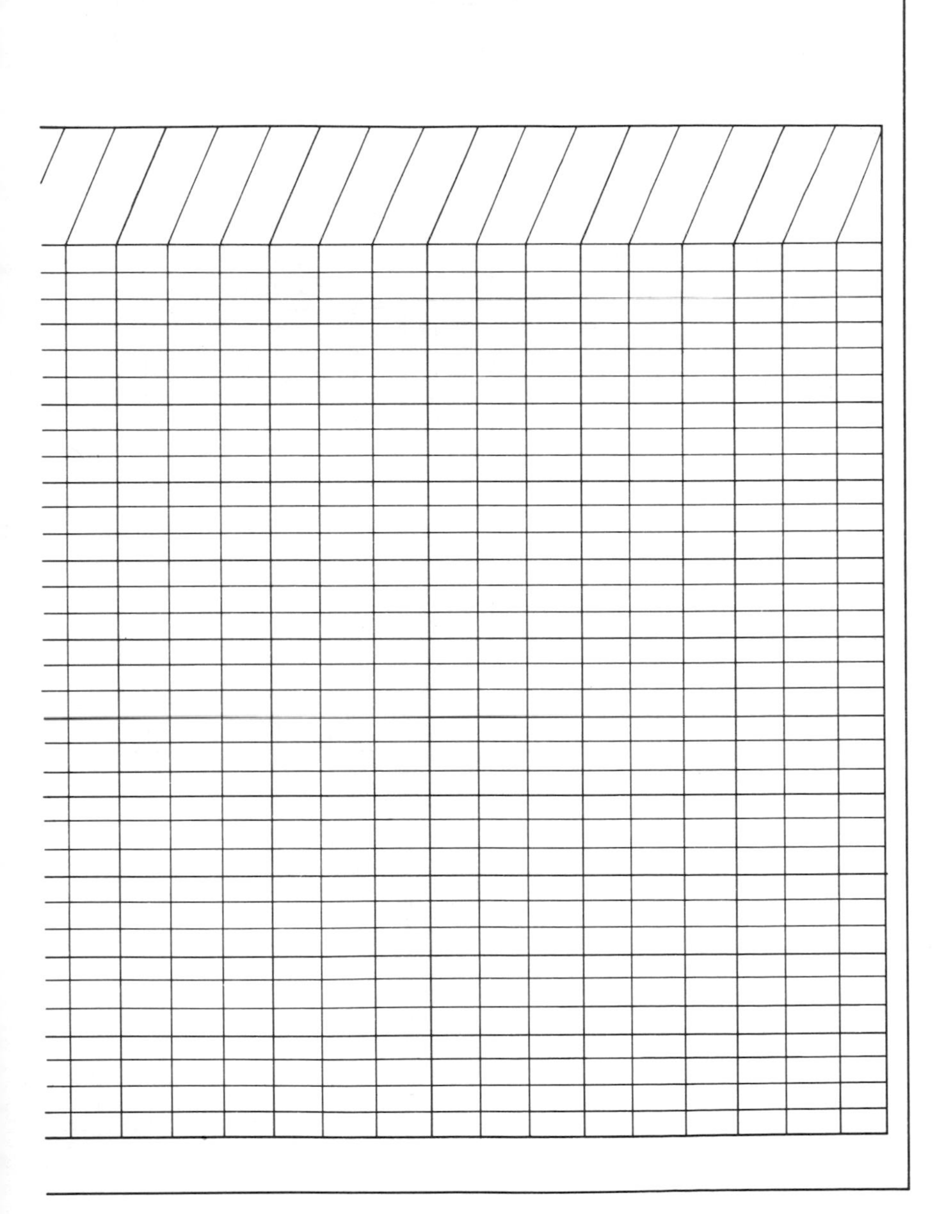

Figure 16.9 (Continued)

BRANCH CIRCUIT SCHEDULE

ESTIMATE NO.__________

JOB__________ LOCATION__________ SHEET NO.__________

OWNER__________ ADDRESS__________ OF__________SHEETS

WORK__________ SCALE__________ ESTIMATED BY__________ DATE__________

FLOOR	OUTLETS												PIPE ENTRANCES						CONDUIT AND WIRE				
	CEIL.	BRKT.	FLR.	P.R.	SWITCHES								CEILING			WALL							
					S.P.	3–W.	4–W.						1/2″	3/4″	1″	1/2″	3/4″	1″					

	WIRE SUMMARY			OUTLETS	
	No. 14	No. 12			
1/2" CONDUIT				CEILING	
3/4" "				BRACKET	
1" "				FLOOR	
				PLUG REC.	
				SWITCH S.P.	
ODD WIRE				" 3–W.	
OUTLETS				" 4–W.	
CIRCUITS					
ALLOWANCE					
TOTAL					

Form E–9

Figure 16.10. Branch circuit schedule.

FEEDER SCHEDULE

JOB__________ DATE__________ ESTIMATE NO.__________

__________ ESTIMATED BY__________ SHEET NO. ______ OF ______ SHEETS

FEEDER NO.		FROM	TO	CONDUIT					WIRE					
				SIZE	LENGTH	LS.	TERM.	BENDS	NO.WIRES	SIZE	LENGTH			

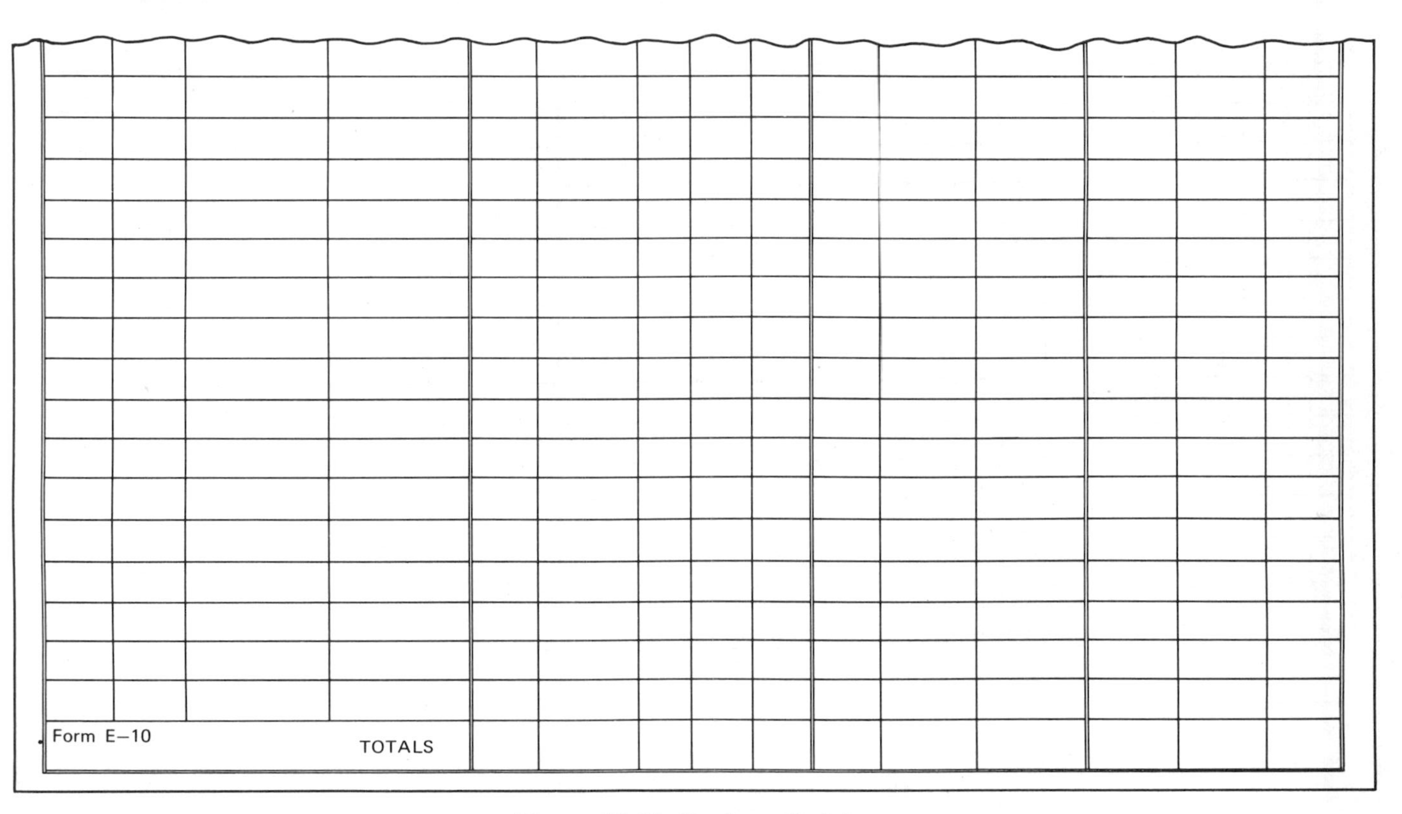

Figure 16.11. Feeder schedule.

CONDUIT AND WIRE SUMMARY

ESTIMATE NO. ____________

JOB ____________ SHEET NO. ____________

ESTIMATED BY ____________ DATE ____________ OF ____________ SHEETS

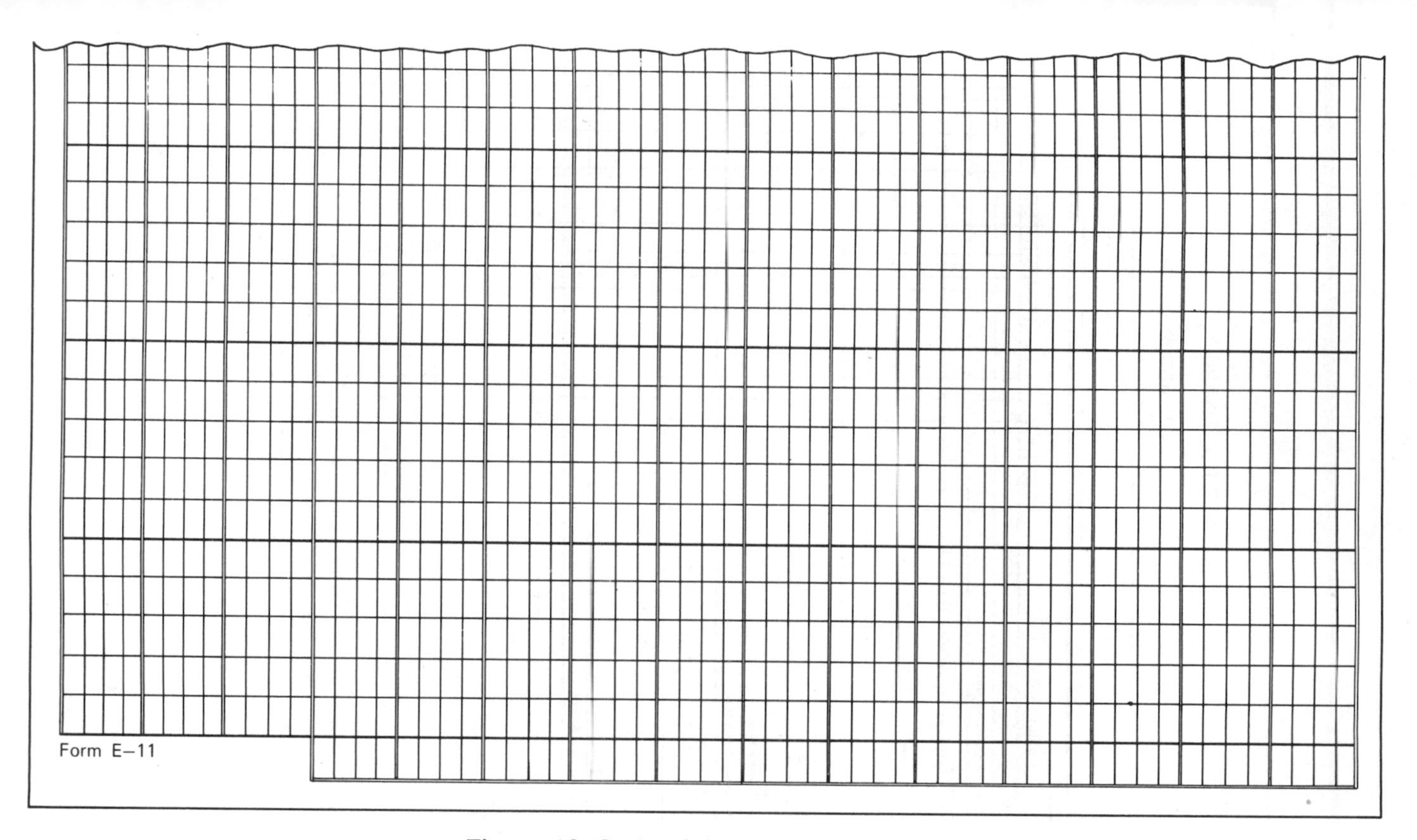

Figure 16.12. Conduit and wire summary.

HOUSE WIRING SUMMARY

JOB ______________________________ ESTIMATED BY ______________ PAGE ________

______________________________ CHECKED BY ______________ OF ______ PAGES

____

____

DESCRIPTION	OUTLETS MATERIAL				LABOR		
	QUANTITY	UNIT PRICE	PER	AMOUNT	UNIT	PER	AMOUNT
Ceiling or Wall Light Outlet							
Recessed Box with Outlet Box Att.							
Single Pole Switch							
Three Way Switch							
Four Way Switch							
Weatherproof Switch, Add for Each							
Dimmer Switch							
Door Switch							
Pilot Light on Switch, Add for Each							
Duplex Receptacle 2 Wire							
Duplex Receptacle 3 Wire Grd.							
Duplex Receptacle Split Wired or Switched							
Duplex Receptalce Weatherproof with G.F.I. Recept. or Brkr.							
Floor Receptacle							
Clock Hanger Receptacle							
T.V. or Radio Receptacle							
Telephone Outlet, Box & Plate							
Disposal							
Kitchen Exhaust Fan (or Hood)							
Dishwasher							
Electric Range with Recept.							
Electric Oven							
Electric Cook Top							
Electric Clothes Dryer with Recept.							
Electric Water Heater							
Electric Water Pump ☐ 110 V. ☐ 220 V.							
Furnace Oil or Gas Burner							
Thermostat							
Furnace Blower w/switch							

Electric Heat Units Recessed																		
Electric Heat Units Baseboard																		
Electric Heat Units Wall Mount																		
Electric L.V. Thermostats																		
Electric Highvoltage Thermostats																		
Receptacle Insert Sections																		
Bath Room Heater																		
Bath Room Exhaust Fan																		
Receptacle 20 Amp. 120 V. Grd.																		
Receptacle 20 Amp. 240 V. Grd.																		
Receptacle 30 Amp. Air Cond., etc.																		
Low Voltage Switch – Single Relay																		
Relay & 1 Flush Switch																		
Added Flush Switch																		
Relay & 1 Surface Switch																		
Added Surface Switch																		
Master Switch																		
Gang Box Relay																		
Relay, Install & Connected																		
Flush Switch, Install																		
Surface Switch, Install																		
Master Switch Install																		
Transfomer Mounted & Connect.																		
Electronic Air Cleaner or Humidifer																		
Bell Wiring w/Transformer																		
Smoke Detectors																		
Service: 60 Amp. – Circuit																		
100 Amp. – Circuit																		
150 Amp. – Circuit																		
200 Amp. – Circuit																		
Fuse–Center or Panels																		
Electric Permit/Inspection Fee																		
Temporary Service Set Up																		
Underground Service																		
		TOTAL																

Figure 16.13. House wiring summary.

FORM E-23

MINNESOTA ELECTRICAL ASS'N
3100 Humboldt Ave S
MINNEAPOLIS MN 55408

HOUSE WIRING RECAP

JOB ______________________ ESTIMATED BY ______________ PAGE ________

______________________ CHECKED BY ______________ OF ________ PAGES

DESCRIPTION	OUTLETS – MATERIAL				ACCOUNTING		
	QUANTITY	UNIT PRICE	PER	AMOUNT	UNIT	PER	AMOUNT
Ceiling or Wall Light Outlet							
Recessed Fixture Pull Chain ☐ Switch ☐							
Single Pole Switch							
Three Way Switch							
Four Way Switch							
Weatherproof Switch, Add for Each							
Dimmer Switch Single Pole ☐ 3 Way ☐							
Outside Floods or Post Lights							
Electronic Air Cleaner — Humidifier							
Central Air Conditioner							
Duplex Receptacle 3 Wire Grd. 15 Amp							
Duplex Receptacle 3 Wire Grd. 20 Amp							
Duplex Receptacle Split Wired or Switched							
Duplex Recept. Weatherproof w/G.F.I. Recept. or Brkr.							
Smoke Detector							

Chime Wiring with _____ Button(s)																		
Laundry Receptacle																		
Mercury Vapor Fixture																		
Disposal																		
Kitchen Exhaust Fan (or Hood)																		
Dishwasher																		
Electric Range — With Recept.																		
Electric Oven																		
Electric Cook Top																		
Electric Clothes Dryer with Recept.																		
Electric Water Heater																		
Electric Water Pump 110 V. ☐ 220 V. ☐																		
Furnace Oil or Gas																		
Thermostat																		
Bathroom Fan ☐ Fan Light ☐																		
100 Amp Service																		
150 Amp Service																		
200 Amp Service																		
Temporary Service																		
Electrical Permit Fee																		
TOTALS																		

Figure 16.14. House wiring recap.

smaller contractors also. The basic concept of the automatic estimating system is as follows:

As the material take-off is being accomplished through the use of an automatic electronic scaling and counting probes or by manual input directly to the key board, quantities are fed into the system which is followed by a simplified identification code number.

From that simple step, internally, a minicomputer calculates and stores the basic material, required labor hours, and material cost. At the same time, it automatically computes any by-product materials, such as hangers, terminations, connectors, and couplings, plus labor hours and material cost for them. As the take-off progresses, all the quantities, man-hours, and costs are summarized and recapped toward a final detailed summary, formatted as the contractor desires.

The most advantageous phase of estimating as applied to the system is branch circuitry, plus feeder schedules versus panels, fixtures, disconnects, and so on, simply because branch circuitry and feeders have much more uniformity and less deviation than the other items.

An automatic estimating system also allows for automatic pricing and labor computation. Yet it has the ability to change prices as required in a simple automatic fashion with no knowledge of programming required.

The final printout of most systems is in triplicate and can then be distributed to estimating and management for bidding purposes, to warehouse, and to accounting for use in job and progress and cost control.

In addition, customized programming allows use of the contractor's own labor units, with the ability to adjust them in accordance with various job conditions. In short, the customized programs reflect the individual contractor's bidding philosophy and methods.

17

Wiring System Take-off

The National Electrical Contractor's Association in Washington, D.C. is noted for having developed the first *practical* manual of labor units. To quote from the 1929 *NECA Estimating Manual*:

> The experience of many estimators over a term of years has conclusively demonstrated that, except in certain special cases, there is only one method which can be considered satisfactory for estimating the cost of electrical construction work. This one acceptable method is to first take off a complete list of the material required for the installation, and then to compute the labor cost by applying the proper labor units to the material items.

The preceding statement has been proven over and over again in the electrical industry and the principle indicated in the statement has become the basis of sound estimating procedures even today. While certain standardized operations can be priced on the unit-price method, all nonstandardized jobs must be prepared by making a complete material take-off, and then applying appropriate labor units.

17.1 Wiring System Take-off

A quantity survey or wiring system take-off consists of counting all the outlets by type (switch, receptacle, etc.), lighting fixtures by type, panelboards and similar equipment and the measurement of all branch circuit runs, feeders, services, and the like. These quantities are entered in their appropriate spaces on a material take-off form such as the one discussed in Chapter 16.

Some estimators make a very detailed material take-off, listing all branch circuits separately and including such small items as wire nuts and fastenings. Others will take off the major items of material for an entire building listing only the different types of materials separately. In most cases, however, the estimator will measure each feeder, service conductors, and other heavy power raceways separately, measure all the No. 12 AWG wire raceways as a group; all the No. 10 AWG raceways as a group; and so on.

It really doesn't matter which procedure is used as long as the estimator has sufficient information from which to make a complete list of all materials required to complete the installation. With this list he can apply appropriate labor units and have a means for pricing and ordering the materials.

A typical material take-off begins with the counting of all duplex receptacles shown on the working drawings and the number entered in a space on the pricing sheet (see Chapter 16). The estimator continues counting all lighting switches, junction boxes (requiring standard outlet boxes), fire alarm, telephone outlets, and so on, listing each specific type separately.

The counting of all lighting fixtures by type will normally come next; each type and type group listed separately. All panelboards, safety switches, pull boxes, and related equipment comes next along with any special equipment and appliances.

With all of the major items out of the way, the estimator will continue the take-off by measuring all branch circuit runs and listing all wire sizes and conduit sizes separately; the measurement of all feeder, service, and other heavy power runs; accounting for all special raceways, bus duct, and special power equipment.

The actual mechanics of a material take-off procedure is comparatively simple, and will become almost routine in a very short time. The sooner the estimating procedures become routine, the more rapidly and accurately the estimator can make the take-off.

Basically, the estimator will hold some type of counter or tabulator in his left hand and will begin checking off the various outlets with a pencil held in the right hand (for a right-handed person). Each time an outlet is checked off, the counter or tabulator is tripped, recording the counted outlet. When all of one type of outlets have been accounted for, the tabulator scale is read and this number entered on the take-off form. The estimator then continues on to the next outlet type and counts all of these in the same manner.

When all the outlets, lighting fixtures, panelboards, and so on have been counted and entered in their appropriate spaces in the take-off form, the estimator will use some type of mechanical device, like a rotometer, to measure the various circuit runs. On straight runs such as feeders, an architect's scale may be used, but due to the short length of these scales they are often awkward to use. A better device is the tape measure which is calibrated in various scales; that is, ¼ inch = 1 foot, ⅛ inch = 1 foot, and so on.

While measuring the circuit runs, the ones measured should be marked in some way so as not to measure a run more than once. A colored pencil is fine for this marking.

The estimator should remember that the circuit lines on floor plans represent only the horizontal portion of the various runs. Therefore, in order to accumulate the vertical runs in the system, the estimator should acquire a scaled section of the building showing the various floors and ceiling heights. Then the mounting heights of duplex receptacles, wall switches, telephone outlets, above-counter outlets, wall-mounted lighting fixtures, and so on should be scaled on this building section. This can be done by merely drawing a straight line at the appropriate heights and then identifying each line with the appropriate symbol.

On conduit or circuit runs, whenever a point is reached where there is a vertical section of the run, the rotometer or other measuring device is run over the proper vertical distance on the scaled cross-section. This will continually accumulate the vertical distances along with the horizontal distances on the floor plans.

There are several other aids that will help the estimator make an accurate take-off. One is to use different colors of pencils in checking off runs of conduit as they are measured—a different color for each size conduit. For example, black could be used to indicate ½ inch conduit, blue to indicate ¾ inch, and so forth. Sizes of conduit as well as terminals (connectors) can then be easily counted and/or determined.

17.2 Listing the Material

While taking off the various electrical components from working drawings, the estimator must list the items on pricing sheet forms so that prices may be obtained for the various items and labor units may be applied and extended. To help make this operation easier for both the estimator and the purchasing agent, the listings should be done in an orderly sequence on the pricing sheet as shown in Figure 17.1.

ESTIMATE SHEET

JOB Warrington Fire Dept. PAGE 1

Route 15 OF 6 PAGES

Warrington Virginia

ESTIMATED BY CHECKED BY DATE

	DESCRIPTION	MATERIAL				LABOR			
		QUANTITY	UNIT PRICE	PER	AMOUNT	UNIT	PER	AMOUNT	
1	No. 12 TW wire (white)	10,500′							1
2	No. 12 TW wire (black)	10,500′							2
3	No. 12 TW wire (red)	6,000′							3
4	No. 12 TW wire (blue)	3,800′							4
5	No. 10 TW wire (white)	1,750′							5
6	No. 10 TW wire (black)	1,750′							6
7	4″ Sq. × 1½″ deep Gal. boxes (¾″ KO's)	425							7
8									8
9	etc.								9
10									10
11									11

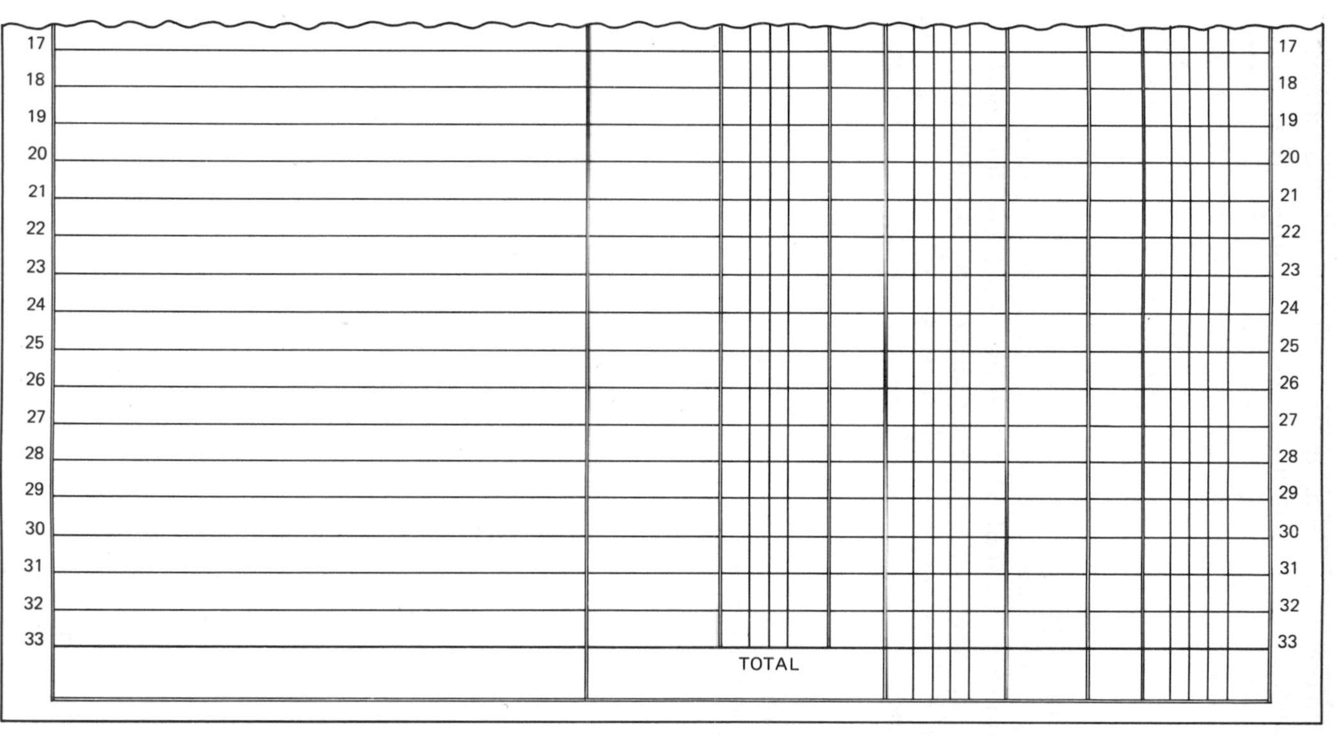

Figure 17.1. The listing of materials should be done in an orderly sequence.

When the take-off has been performed properly, the estimator will immediately have two items of valuable information: (1) a brief description of each outlet, conduit or cable run, component, equipment, and so on, and (2) the quantity of each item listed. From these descriptions, the estimator can determine the exact quantity of materials and the necessary labor hours to completely install the system, provided the estimator has a good knowlege of actual electrical installation and building installations.

Certain items of the take-off sheet must be converted in order to account for all items of material. For example, the take-off sheet may indicate a total of 6500 feet of ¾-inch conduit containing four No. 12 AWG solid copper conductors. This total should then be further broken down so the correct amount of materials may be ordered, as well as miscellaneous items like pipe hangers and fastenings.

To illustrate a typical listing, assume that the 6500 feet of ¾-inch conduit with four No. 12 AWG conductors is to be installed in a concrete slab with only 18 inches of exposed pipe running up to each outlet. Further assume that this total run of conduit will feed 220 duplex receptacles and 89 wall switches. With this information, the material will be listed something like the following:

6500′ ¾″ rigid conduit
6500′ No. 12 TW building wire (white)
6500′ No. 12 TW building wire (black)
6500′ No. 12 TW building wire (blue)
6500′ No. 12 TW building wire (red)
178 ¾″ rigid pipe straps (one-hole)
178 No. 10 sheet metal screws with corresponding plastic or fiber anchors for concrete block walls

The conversion of the duplex receptacle outlets (220 total) will be:

220 4″ sq. × 1½″ deep gal. outlet boxes with ¾″ KOs
700 ¾″ locknuts
440 No. 10 sheet metal screws and corresponding anchors
220 4″ sq. single-gang plaster rings, ½″ deep
220 Duplex receptacles (manufacturer and cat. no.)
220 Stainless steel duplex recpt. plates
220 grounding screws with bare ground wire attached

Such incidentals as fastenings, hangers, and wire connectors are rarely indicated on the drawings. Therefore the estimator will have to make a calculated determination based upon his understanding of the project's requirements, past experience with other projects, and the use of good judgment. Still, when the estimated fastenings and hangers are calculated along with each run of conduit, it is possible to make a reasonably accurate list of the material items required.

When the estimator has finished compiling a complete material and labor operation list, he has a complete record of the descriptive and layout data contained in the specifications and shown on the working drawings along with the required labor and material items.

There is really no easy way to accomplish this result, but experienced estimators produce very accurate estimates, rarely omitting important items, when a systematic estimating method is used. These same estimators also know that when the basic principles of electrical estimating are known, these principles can be adapted to almost any type of electrical installation.

The extent of the detail in which estimators list items of material can vary to suit the contractor's particular estimating method, but experience has proven that the more detailed, the better.

17.3 Basis of Applying Labor Units and Pricing Material

To properly apply labor units to the extended material units requires a great deal more than the use of published labor unit books. By the same token, the pricing of materials also requires an intelligent analysis of the quotations by suppliers or price services.

Many factors govern the accuracy of labor unit and material data including: (1) whether the data is up-to-date; (2) whether the data is correct as applied to the particular project, or if contingency allowances have to be made; (3) whether the data is based upon the same quantities and types of materials as listed in the pricing sheets.

Most pricing forms are arranged so that the listed item of materials also identifies the labor operation required to install it. However, in some cases, certain labor allowances will be required to completely install a given piece of equipment for which there is no material required. Such a situation would naturally require that extra labor be allowed, but this operation is often overlooked—if no material items are listed.

SEPTEMBER 10, 1974
CANCEL JULY 23, 1974
PRICE INCREASE

RIGID ALUMINUM CONDUIT
Couplings & Elbows

9-10-74

ALUMINUM COUPLINGS FOR RIGID CONDUIT
Prices shown as published by CONDUIT PIPE - Weights by KAISER

Size	Weight Per 100 Pieces	Net Per 100 Pieces	UNIT PRICE
$\frac{1}{2}$	6.1	35.00	$ .63 B
$\frac{3}{4}$	9.1	51.67	.93 B
1	12.5	60.00	1.08 B
$1\frac{1}{4}$	18.9	73.33	1.32 B
$1\frac{1}{2}$	23.3	88.33	1.59 B
2	34.6	120.00	2.16 B
$2\frac{1}{2}$	68.3	253.33	4.05 C
3	91.4	368.33	5.90 C
$3\frac{1}{2}$	108.0	551.67	8.80 C
4	142.0	600.00	9.60 C
5	241.9	1435.00	23.00 C
6	321.0	1730.00	27.50 C

ALUMINUM ELBOWS FOR RIGID CONDUIT
Prices shown as published by CONDUIT PIPE - Weights by KAISER

Size	Weight Per 100 Pieces	Net Per 100 Pieces	UNIT PRICE Margin C
1/2	29.0	86.67	$ 1.39
3/4	43.0	131.67	2.11
1	71.0	186.67	3.00
1 1/4	110.00	270.00	4.30
1 1/2	153.0	306.67	4.90
2	249.0	468.33	7.50
2 1/2	437.0	871.67	14.00
3	767.0	1400.00	22.50
3 1/2	1036.0	2521.67	40.50
4	1228.0	2923.33	47.00
5	2490.0	8991.67	144.00
6	3850.0	12215.00	195.50

Our Suggested Resale (Unit) Prices above Include Costs, Estimated Overhead anc a Fair Profit.

MQ

Figure 17.2. Example of pricing sheet from a pricing service offered by National Price Service, 4525 West 10 Street, Cleveland, Ohio 44135.

The majority of electrical contractors list the cost of material and labor operations separately, and then price each, when compiling estimates for lump-sum, firm-contract electrical construction projects. The advantages of this method are many, but the more important ones include:

1. Prime cost of the overall project (as well as individual sections) can easily be determined.
2. Basic costs are more easily adapted to progressive job cost analysis.
3. Different percentages of markup for different types of material and equipment can be applied if necessary or desirable.

17.4 Pricing Material

Material prices are available from several different sources, but on the majority of electrical construction bids, the estimator usually obtains a firm quotation from one or more electrical suppliers on the major items like lighting fixtures (with lamps), panelboards, service equipment, wire and cable, and conduit.

Current prices are also available from the manufacturers, but often electrical suppliers will compete with one another to obtain an order for a cream project. Therefore, if the estimator uses the published book projected, especially on the larger coveted projects, he will almost always be high on his estimate, and probably never get a job.

There are certain recommended procedures and precautionary measures the electrical contractor should follow when obtaining quotations from electrical suppliers. The first is to make requests for quotations on special materials as early as possible, although electrical suppliers often wait until the last few minutes prior to bid openings before they will give out a quotation. This, of course, is done to prevent another supplier from underbidding them.

When the quotation is received, the electrical contractor should check over the list of items on the quotation carefully. Electrical suppliers do not normally guarantee that the items will meet with the project's specifications, nor will they take any responsibility for errors. Substitutions are common these days, and it is the contractor's responsibility to make certain that all items quoted will meet with the architect's or engineer's specifications. The contractor should also check the quantities of the quotation against those obtained from the take-off to make sure they are correct.

Whenever possible, the electrical contractor should obtain a guarantee of the quoted price for a definite period of time. Most suppliers will stand by their quotes for approximately 30 days. But what happens if it takes six weeks to award the contract? There is a good chance that the material quote has raised in price making the contractor who gets the project pay higher prices for the material than he originally figured in his bid. This means that he will make less profit and it could even result in a loss. Therefore, the electrical contractor should try to determine exactly when the job will be awarded (this is not necessarily the date of the bid opening) and then obtain a guaranteed price from his suppliers until that date.

The electrical contractor should further determine if the quotation for material prices used in his bid is actually valid. Is he absolutely certain that this supplier will deliver the required quantities of material at the time and place required by the job?

Many electrical contractors feel that shopping around for a lower price than the one used for his bid is good business practice. However, many such contractors are often faced with the refusal of some of the suppliers to furnish them with quotations. Experienced contractors have found it best to honor the lowest quotation given from suppliers which is used for the basis of pricing material for the bid. They seldom shop around after the contract is awarded to see if they can chisel some of the prices down.

18

Applying Labor Units

A labor unit is a time figure indicating the time required to install, connect, or otherwise make usable a given item of material or a given labor operation.

These units are used by the majority of electrical contractors who must quote a firm lump-sum price to obtain electrical construction work, and the units are normally based upon man-hours or a percentage of a man-hour. For example, 1.50 man-hours indicates 1½ man-hours; that is, the labor will normally take one man 1½ hours to perform the installation.

Labor units are applied to each item of material and then extended and totaled to give the total man-hours required to complete the project. The value of the labor in dollars-and-cents is then determined by multiplying the total man-hours computed by the *average* hourly rate of pay for the electrical workers.

There are several manuals of labor units available to electrical contractors either on a purchase or lease basis. However, there is more to using these units than just copying them out of a manual; a more detailed discussion follows. Furthermore, each electrical contractor must select an estimation method (including choice of labor-unit manual) that best suits his individual needs.

18.1 General Considerations for Applying Labor Units

A separate unit of labor should be provided for the installation of each item of material or labor operation performed. This unit should be broken down further to apply to varying working conditions. For example, if a labor unit is given for installing 100 feet of rigid conduit at ground level, it stands to reason that installing the same amount of conduit 20 feet above the floor would require more man-hours; scaffolding would have to be set up and moved into the areas and the men would be required to spend more time lifting materials to the scaffold platform.

There are several other conditions that will affect the labor operation and the electrical contractor must give consideration to all of them in preparing any and all bids. Some of these conditions include:

1. The type of building construction.
2. Height of the installation above normal working areas.
3. The weight of the material or equipment.
4. Performance of the general contractor.
5. The availability and caliber of electrical workers.
6. Whether the wiring installed is concealed or exposed.
7. Whether the installation is installed in new or existing buildings.

The basic labor operation for any electrical installation must take into consideration several factors which are often overlooked by the inexperienced contractor or estimator. For example, the labor unit must include layout instructions, handling the material, the actual installation of the material, coffee breaks, visits to the rest room, and so on. Most labor-unit manuals, however, include all of these factors in their individual units. If not, the contractor must include allowances to cover these items. From this, we see that the amount of time that it takes an electrician to install a given item, at first glance, may not be an adequate basis for determining an accurate labor unit.

18.2 Factors Affecting Labor Units

The accuracy of labor units depends to a great extent upon the type of electrical contracting organization and the type of work performed. For example, one popular manual of labor units (offered by one of the electrical contractors' associations) is reported to be very accurate for electrical installations on average-size commercial projects when using union labor. This same system, however, seems to be about 20% high on smaller commercial projects, especially when permanent well-trained electricians are used, and too low on large industrial projects.

Another manual of labor units distributed by the Minnesota Electrical Contractors Association is reported to be about 18% high for most contractors doing average commercial projects. Therefore, it seems that regardless of the method used, no labor-pricing method is exactly correct unless it is properly applied; the estimator still has to use good common sense.

On the other hand, when soundly determined labor units are properly applied, it is surprising how close the actual labor on a given project compares with the estimated labor.

The productivity of labor will more than likely vary from locality to locality, and the electrical contractor or his estimator must try to determine, if at all possible, the efficiency of the electricians he will use, prior to bidding the job. When this is not possible, good judgment must enter into the picture, and appropriate factors must be applied.

Such items as the general contractor's performance, local union's cooperations, and promptness in paying all have a bearing on the labor expended. For this reason, it is not possible to have one set of labor units that will be correct for all types of installations under all conditions. Labor units are good to have as a guide, but experienced contractors know that any published set of labor units will have to be modified to suit particular situations.

While it is not possible to list all of the variable factors affecting labor costs (and in turn, labor units), most of them can be classed under the following general headings:

1. Nonproductive labor.
2. The ability and attitude of the general contractor.
3. The ability and attitude of other subcontractors.
4. Type of building.
5. Working conditions.
6. The ability and efficiency of the electrical contractor.

18.3 Application of Labor Units

As briefly pointed out in the previous section, the electrical contractor or his estimator is continually faced with having to use good judgment (and educated guesses) when dealing with labor units. At first, the selection of the proper labor units may seem to be a difficult task, but after some experience in the field, the experienced estimator will be able to readily choose the labor units most applicable to the particular project or portion of the project.

The first step in arriving at the most accurate total estimated labor for a given job is to take off and list the material items on the pricing sheets, segregated in accordance with the installation and building conditions. On a larger and more complex job the different categories applicable to each type of material can be expanded in line with the different specific installation conditions.

The second step is to apply the labor unit specifically related to that particular installation condition for the size and type of material involved, depending on the extent of segregated listing of the materials and the extent of segregation of the available labor data. However, there is no point in listing the materials on a segregated basis if segregated labor data is not available, or the estimator does not adjust the existing data to account for the specific conditions. Anything less than a segregated listing of the materials in accordance with the varying installation conditions and the application of related labor data reduces the accuracy of the total estimated labor.

There are many variations in the types and sizes of the basic materials which affect the amount of labor expended in their installation. However, by keeping in mind the characteristic labor operations required to put the basic materials in place, the estimator can more readily visualize the labor requirements for all variations in the types and sizes of the materials and select the applicable labor units.

The unusual installation for which there appears to be no applicable specific labor units can usually be reduced to a series of component individual labor operations for which an actual or comparable labor unit may be available.

When no published or comparable cost data are available, the estimator has no alternative but to make allowances based upon previous general experience. The installation should be broken down into the greatest possible number of component operations or items of material. The estimator then reflects upon the probable amount of time it should take an electrician to perform each bit of work. An experienced estimator will not usually misjudge the time of each item of work to any great extent, and the total of the labor allowed for all the component items will be fairly close to the labor expended in most instances. A novice estimator, however, should proceed with caution and discuss the problem with an experienced foreman or superintendent.

In choosing the proper labor units, the estimator must be

guided by only one principle and that is his responsibility to include in the estimate the amount of labor which will have been expended when the job is completed. The units must be chosen with some degree of intelligent analysis and not used blindly just because they happen to appear on a printed sheet.

Once the choice of labor unit has been made, the mechanics of labor unit entry are merely the copying of the appropriate unit from the manual or data sheet and entering it in the labor unit column on the pricing sheet and on the line opposite the proper item of material or labor operation.

After all of the labor units have been applied on the pricing sheet, they should be extended and totaled. This operation involves little more than elementary mathematics, but many errors are easily made in such computation and the estimator should therefore be extremely careful at this point. One decimal point in the wrong place can mean the difference between a profit and a loss on a project.

It is recommended that a good electronic calculator be used in making all extentions and totals. Besides saving a great deal of time, as well as fatigue, they greatly reduce the element of human error.

In using the calculator, the simplest procedure is to place the calculator on the left (for a right-handed person) with the pricing sheet to the right. As the estimator uses a rule as a guide to line up the figures on the sheet, he operates the calculator with his left hand and then writes the extensions, figures, etc. with his right hand. However, some right-handed persons cannot get use to using the calculator with their left hand; so an extra step is necessary; that is, using the left hand to line up the figures with the rule while operating the calculator with the right hand, and finally using the right hand to write in the figures.

To help keep the figures on the proper line, a pencil line should be drawn across the line space on all lines that do not contain figures.

No bid should ever be turned in without checking over all figures at least once. Preferably, the person making the initial take-off should check through the figures first, and then someone else should quickly check them over. One method of checking column totals is to first add them from top to bottom and then from bottom to top. In any case, sufficient time should be allowed for this checking, as errors often result from hasty last-minute efforts to complete an estimate to meet a specific bid time.

19

Summarizing the Estimate

Summarizing the estimate is the final accumulation of all estimated costs—labor, material, job factors, direct cost, overhead and profit—and the determination of the final quotation. It is one of the most important steps in preparing the estimate because one mistake in the final summarizing can affect all of the accuracy with which the previous steps have been handled.

19.1 Summarizing in General

A typical bid summary sheet appears in Figure 19.1 and includes the following basic sections or groups of cost data:

1. Description of the project
2. Cost of listed material and labor
3. Nonproductive labor
4. Direct job expenses
5. Taxes, bonds, etc.
6. Overhead
7. Profit

A form such as the one just described serves as a sound guide to accurate summarizing of the estimate. Besides providing titled spaces for the entry of all necessary cost figures, they also serve as a valuable check list to eliminate omissions.

BID SUMMARY SHEET

JOB ______________________________ SHEET NO. ____ OF ____ SHEETS

____

____

ESTIMATED BY ____________ CHECKED BY ____________ DATE ____________

SHEET NO.	DIVISION	MATERIAL--DOLLARS	LABOR--HOURS

NON-PRODUCTIVE LABOR	HOURS
Handling Material	
Superintendent	
Traveling Time and Lost Time	
Job Clerk	
TOTAL (A)	

JOB EXPENSE	DOLLARS
Tools, Scaffolds	
Pro Rata Charges	
Insurance, Public Liability, Etc.	
Cutting, Patching, Painting	
Watchman	
Telephone	
Drawings	
Inspection and Permit Fees	
License	
Storage	
Freight, Express and Cartage	
Transportation	
Board ___ Men ___ Weeks At ______	
TOTAL (B)	

Miscellaneous Material and Labor		
Non-Productive Labor		(A)
TOTALS – MATERIAL (C) & LABOR (D)		
(D) ______ Hours Labor @ ______		
(D) ______ Hours Labor @ ______		
(D) ______ Hours Labor @ ______		
Taxes: Soc. Sec. ______ Unemp. ______		
Workmen's Compensation Insurance		
LABOR COST GROSS TOTAL		
Job Expense (B)		
Material Cost (C)		
TOTAL PRIME COST		
______ % Overhead		
TOTAL NET COST		
______ % Profit		
Selling Price Without Bond		
Bond		
Selling Price With Bond		
PRICE QUOTED		

To Avoid Errors, Check List on Reverse Side   Let's Upgrade Our Electrical Industry

Figure 19.1. Bid summary form (front side).

CHECKING LIST FOR BID SUMMARY AND ESTIMATE SHEETS

To avoid or catch some of the common errors or omissions in making up estimates and bids, check through this list. The short time required may save you money and embarassment. Make sure that all appropriate items are included in your bid properly figured.

CONDUIT, ETC. Sizes	⅜"	½"	¾"	1"	1¼"	1½"	2"	2½"	3"	3½"	4"	
Rigid Heavy												
Thin Wall Steel Tube												
Greenfield												
Conduit Ells												
Couplings												
Connectors												Straps
Lock Nuts—Bushings												Hangers

CONDUIT FITTINGS .... Regular .... Explosion Proof .... Dust Tight .... Water Tight .... Adapters ....

RACEWAYS .... Metal Mold .... Wire Mold .... Window Strip .... Wire Ducts .... Trough .... Under Floor .... Fittings ....

WIRE Single Cond.	16	14	12	10	8	6	4	2	0	2/0	3/0	4/0	250M	300	400	500
Code R. C.																
R W Moisture Resistant																
R H Heat Resistant																
R P Performance Type																
Weatherproof								Racks		Insulators		Connectors		Lags		Bolts ....

CABLES Sizes	12	10	8	6	4	2	0	2/0	3/0	4/0	250M	Larger	1	2	3	4	Conductor
Lead Cover R. C.																	
Underground, Parkway, Etc.																	
Service Entrance						Potheads .... Service Heads .... Junction Boxes .... Splices											
Non-Metallic (Inside)						Cable Hangers .... Straps .... Sill Plates											
ABC - BX (Inside)						Straight Connectors .... Angle Connectors											

CORDS—Lamp .... Heater .... Light Rubber .... Heavy Rubber .... Sockets .... Guards .... Attachments.

PANELS—Distribution .... Light .... Heat .... Power .... Signal .... Cabinets .... Pull Box .... Cutouts .... Meter Board ....

SAFETY SWITCHES—Meter .... Entrance .... Motor .... Oil Switch .... Relays .... Time Switch ....

FUSES—Plug .... Non-Tamperable .... Nec. Cartridge .... One Time .... Renewable .... Extra Sets .... Links ....

GROUND—Rods .... Pipe .... Clamps .... Fittings .... Wire .... Guards .... Hardware .... Exp. Bolts .... Toggle Bolts ....

OUTLET BOXES—All Types and Sizes .... Plaster Rings .... Covers .... Extensions .... Fixture Studs .... Hangers ....

WIRING DEVICES—1 .... 3 .... 4-Way Switches .... Dupl. Recepts. .... Plates .... Spec. Finish .... Special Devices ....

SIGNAL—Chime .... Bells .... Buzzer .... Push Buttons .... Board .... Transformers .... Wire .... Batteries .... Rectifier ....

CLOCKS—TELEPHONE—FIRE—BURGLAR—THERMOSTATS—Conduits .... Wire .... Outlets .... Equipment ....

PUBLIC ADDRESS—RADIO—Equipment .... Speakers .... Conduits .... Outlets .... Shielded Wire .... Antenna ....

MOTORS—Starters .... Controllers .... Push Button Stations .... FANS—Vent. .... Exhaust .... Attic .... Ducts .... Louvres ....

FIXTURES—Typical .... Special .... Recessed .... Show Wind .... Stage .... Outdoor .... Flood .... Guards ....

LAMPS—Mazda .... Silv. Bowl .... Flourescent .... Lumiline .... Pilot .... Neon .... Special ....

AUXILIARIES—Power Factor Correction .... Capaciters .... Starters .... PORCELAIN ....

MISCELLANEOUS—Tape .... Solder .... Screws .... Nails .... | STRUCTURAL IRON, Etc. ....

HAVE YOU DOUBLE CHECKED PLANS FOR ALL DETAILS?

Service Entrances, Location	OUTLETS	All Floors, Basement, Attic.
Switches, Panels, Cabinets	Ceiling	Outside or Underground Wiring.
Feeder Runs, Sub Feeders	Wall Lights	Connecting All Motors, Etc.
Branch Circuits, Wiring System	Switch	Connecting All Equipment to be supplied by others.
Specified Materials & Methods	Floor	Addenda, Changes, Corrections.
Material Supplied by Owner	Receptacles	Assembling and Hanging Fixtures.
Labor Rates and Conditions	Special	

ARE ALL EXTENSIONS FIGURED PROPERLY? Per Each, Doz., 100 (C), Gross, 1000 (M), Feet, Lbs. Etc.
READ BID FORMS CAREFULLY - - - FILL IN AND SIGN PROPERLY
DON'T FORGET CERTIFIED CHECK OR BIDDERS BOND, IF REQUIRED

Figure 19.2. Bid summary form (back side).

A check list is printed on the back of the form shown in Figure 19.1 and appears in Figure 19.2. The form states:

To avoid or catch some of the common errors or omissions in making up estimates and bids, check through this list. The short time required may save you money and embarrassment. Make sure that all appropriate items included in your bid are properly figured.

Note that the checklist reminds the estimator to check all types of conduit fittings, raceways, and so on, including size for ⅜ inch up. Cables are covered as well as cords, panels, switches, outlet boxes, and so on. At the bottom of the form, the question is asked, "Have you double checked plans for all details?" These details include service entrance location, branch circuits, labor rate and conditions, addenda, changes, and corrections.

19.2 Direct Job Expense and Overhead

A thorough understanding of direct job expense and overhead is extremely necessary so that they may be included in the final estimate to defray such costs.

In general, direct job expenses are those costs, in addition to labor and materials, that must be paid for as a direct result of performing the job. In other words, if the job was not performed, these costs would not occur.

Overhead, on the other hand, are all costs that have to paid whether the particular job is being done or not.

An estimate is not complete until all direct job and overhead expenses have been added to the other items entering into the cost of the project. The first problem then is to determine these costs. Direct job expenses are relatively simple to calculate provided the contractor is fair with himself and includes *all* items of expenses directly with the job at hand. Calculating overhead, however, is a different picture altogether. Many contractors take their previous overhead figures and apply them to work that will be performed in the future. This may result in an accurate estimate, but in most cases, the overhead will change in the future, during the performance of the work being bid. Therefore, the electrical contractor should analyze his anticipated future overhead for all jobs being bid at the present.

Another consideration in figuring overhead is the size of the job. It is a known fact that a small job, in most cases, will cause a higher percentage of overhead than a large job. However, the contractor cannot assume that this will always be true, especially in the case of specialized projects.

When the estimate has been completed to the point of adding the overhead, the known data should include cost of materials, cost of labor, and direct job expense. The overhead is then determined by one of the following methods:

1. The overhead for the year may be divided by the gross sales volume for the year, to find the overhead as a percentage of the gross sales volume. This percentage is then applied to the prime cost of the job.
2. The overhead expense for the year may be divided by the total cost of labor, material, and job expenses for the year to find the overhead as percentage of prime cost.

When the estimated annual volume differs from the past annual volume for which an overhead based upon accounting records is obtainable and for some reason the same total dollar cost of overhead expenses must be maintained, the estimator must determine by simple proportion the applicable average overhead percentage of prime cost and apply this percentage in using the job size scale. This is done by estimating the overhead percentage on the basis of past recorded data, adjusted to future volume and size of work. If the job being estimated represents a change in general work pattern or is a special type of job, the estimator must make an intelligent analysis of all the conditions and further adjust the estimated overhead percentage to be applied, including the overall estimate all costs as accurately as possible.

19.3 Completing the Summary

Completing the summary involves only the inclusion of miscellaneous items such as materials and labor. Of course, all major items of material should be included in the appropriate spaces, but such small items as wire nuts, wire-pulling lubrication, tape, and fasteners, come under the category of "miscellaneous."

Few electrical contractors feel that the listing of these miscellaneous items serves any good purpose. Therefore, on most projects, an allowance is made rather than listing and pricing each item individually. This allowance is usually in the form of a lump-sum figure or a percentage gained from experience or an educated guess. As a rule, ½ of 1% is sufficient for all projects except highly specialized ones which will then require some additional thought as to the correct percentage to allow.

Once this figure has been determined, the dollar value should be entered in the appropriate space on the summarizing form.

The electrical contractor will also be required to calculate miscellaneous labor costs on many projects. Such conditions as overtime (required to complete the project within a specified time), labor disputes, or special installations will make the inclusion of extra labor necessary. There is no set rule for calculating this figure exactly—it is a matter of experience and the use of good judgment.

When labor units are used which do not include any or an adequate allowance for job factor, the estimator must determine the percentage by which he estimates the subtotal of basic productive labor which will be affected by the various job factors. Applying this percentage to the net productive man-hours, a total of additional man-hours that will be expended because of the various job factors is calculated and entered in the proper space in the summary.

On jobs that will cause an expenditure of a greater amount of nonproductive labor than allowed for in the basic labor units used, additional allowances for the nonproductive labor must be included in the summary.

On some jobs it may be necessary or desirable for the electrical contractor to subcontract one or more items of work included in the electrical contract, as covered by the scope of work outlined by the plans and specifications, which may not be the work of electrical workers or for which the contractor submitting the estimate may not be equipped or qualified to perform. The value of each subcontracted work must be included in the estimate summary either as a lump-sum quotation given by the other subcontractor or as a calculated total of hourly or other charges.

The subtotals of the dollar value of the labor, material, subcontractors, if any, and direct job expenses are totaled to give the total prime cost. The percentage of applicable estimated overhead determined, as previously discussed in this study unit, is applied and the dollar value of the overhead expense calculated which, added to the prime cost, gives the total gross cost.

The percentage of profit to be included in the estimate is determined either by the contractor himself or after consultation

M-51.1

E S T I M A T I C C O R P O R A T I O N

DATAMATIC

B I D S U M M A R Y S H E E T

JOB NAME CRESENT MFG. CO. DATE 1-1-71

(1) TOTAL MATERIAL (FROM PRICING SHEET) $ 3,203.08

LABOR (FROM PRICING SHEET) 366.43 HRS.

JOB FACTORS — HRS.

(2) TOTAL LABOR @ $ 8.00 /HR, ______ HRS. $ 2,931.44

JOB EXPENSE

SALES TAX 6 % OF LINE 1 $ 192.18

PAYROLL TAX 13 % OF LINE 2 $ 381.09

SUBCONTRACTS $ —

RENTAL OF TOOLS, EQUIP., ETC. $ —

OTHER JOB EXPENSE $ —

(3) TOTAL JOB EXPENSE $ 573.27

(4) PRIME COST (SUM OF LINE 1, 2 & 3) $ 6,707.79

(5) OVERHEAD 15 % OF LINE 4 $ 1,006.17

(6) NET COST (SUM OF LINES 4 & 5) $ 7,713.96

(7) PROFIT 10 % OF LINE 6 $ 771.40

(8) TOTAL COST (SUM OF LINES 6 & 7) $ 8,485.36

(9) BOND $ 86.93

(10) PERMIT $ 36.00

TOTAL SELLING PRICE (SUM OF LINES 8, 9, & 10) $ 8,608.29

PREPARED BY: GR CHECKED BY: JW

Figure 19.3. Completed summary form supplied by Estimatic Corporation, Denver, Colorado 80223.

between the contractor and the estimator, taking into consideration the type and size of job, the character of the competition on the job and the desirability of obtaining the job.

Some contractors desire to apply a flat percentage of profit to all estimates; others vary the percentage in accordance with the factors indicated previously. Some contractors do not use a percentage adder, but determine the dollar value of the profit desired on the basis of a certain amount for each man-day required by the job or by allowing a flat sum of money.

There are certain items of cost which in a true sense are direct job expenses, but against which it may not be desired to assess a profit. Such items may be sales taxes, excise taxes, payment and performance bonds. When any of those items have not been included previously in the estimate, they must be added into the final price.

The total estimated price is calculated by totaling the gross cost, profit and other items. Normally the total estimated price or the nearest even figure is determined to be the amount of the bid. In too many instances when the contractor or estimator becomes uneasy over the competition on the job, the amount of the bid bears little resemblance to the total estimated price.

In some instances some awarding authorities, usually architects, require the bids on a given job or project to be broken down into many segregated items for which unit prices are required. The bid form will usually indicate the quantity of the item or group of work. This gives the owner or architect a means of approving program payments.

Too much emphasis cannot be put on the necessity of including in the summary the proper allowances for direct job expense, job factor, nonproductive labor, labor productivity factor, overhead expenses and profit.

One fundamental point has been stressed. The cost factors included in the summary must be determined for each individual job or project and reflect the individual cost characteristics of each job.

Any estimate properly summarized will more nearly provide enough income from that job to pay for all costs, both direct and indirect, caused by that job than if the final price is established on a hit-or-miss basis. When each job proves to be reasonably profitable, the entire business operation is successful.

Glossary

accessible (as applied to wiring methods): capable of being removed or exposed without damaging the building structure or finish, or not permanently closed in by the structure or finish of the building.

accessible (as applied to equipment): admitting close approach because not guarded by locked doors, elevation, or other effective means.

aggregate: inert material mixed with cement and water to produce concrete.

alternating current: current (ac) which reverses direction rapidly, flowing back and forth in the system with regularity. This reversal of current is due to reversal of voltage which occurs at the same frequency. In alternating current, any one wire is first positive, then negative, then positive and so on.

alternator: an electric generator designed to supply alternating current. Some types have a revolving armature and other types a revolving field.

ampacity: current-carrying capacity expressed in amperes.

ampere: the unit of measurement for electric current. It represents the rate at which current flows through a resistance of one ohm by a pressure of one volt.

amplitude: the maximum instantaneous value of an alternating voltage or current. It is measured in either the positive or negative direction.

appliance: utilization equipment, generally equipment other than industrial, normally built in standardized sizes or types and installed or connected as a unit to perform one or more functions, such as clothes washing, air conditioning, food mixing, deep frying.

appliance, fixed: an appliance that is fastened or otherwise secured at a specific location.

appliance, portable: an appliance that is actually moved or can easily be moved from one place to another in normal use.

appliance, stationary: an appliance that is not easily moved from one place to another in normal use.

approved: acceptable to the authority enforcing the Code.

attachment plug (plug cap) (cap): a device that, upon insertion in a receptacle, establishes a connection between the conductors of the attached flexible cord and the conductors connected permanently to the receptacle.

automatic: self-acting, operating by its own mechanism when actuated by some impersonal influence, such as a change in current strength, pressure, temperature, or mechanical configuration.

backfill: loose earth placed outside foundation walls for filling and grading.

ballast: an electrical circuit component used with fluorescent lamps to provide the necessary voltage for striking the mercury arc and then to limit the amount of current flowing through the lamp.

bearing plate: steel plate placed under one end of a beam or truss for load distribution.

bearing wall: wall supporting a load other than its own weight.

bench mark: point of reference from which measurements are made.

bonding jumper: a reliable conductor used to ensure the required electrical conductivity between metal parts required to be electrically connected.

branch circuit: that portion of a wiring system extending beyond the final overcurrent device protecting the circuit.

branch circuit, appliance: a circuit supplying energy to one or more outlets to which appliances are to be connected; such circuits have no permanently connected lighting fixtures that are not a part of an appliance.

branch circuit, general purpose: a branch circuit that supplies a number of outlets for lighting and appliances.

branch circuit, individual: a branch circuit that supplies only one piece of utilization equipment.

bridging: system of bracing between floor beams to distribute floor load.

building: a structure that stands alone or that is cut off from adjoining structures by fire walls with all openings therein protected by approved fire doors.

bus bar: the heavy copper or aluminum bar used on switchboards to carry current.

cabinet: an enclosure designed for either surface or flush mounting and provided with a frame, mat, or trim in which swinging doors are hung.

capacitor or condenser: an electrical device that causes the current to lead the voltage, opposite in effect to inductive reactance. They are used to neutralize the objectional effect of lagging (inductive reactance) which overloads the power source. Also acts as a low resistance path to ground for currents of radio frequency thus effectively reducing radio disturbance.

cavity wall: wall built of solid masonry units arranged to provide air space within the wall.

chase: recess in inner face of masonry wall providing space for pipes and/or ducts.

circuit breaker: a device designed to open and close a circuit by non-automatic means and to open the circuit automatically on a predetermined overload of current, without injury to itself when properly applied within its rating.

circular mil: the area of a circle 1/1000 inch in diameter. The area of electrical conductors is usually measured in circular mils; that is, 500,000 circular mils (500MCM), etc.

coaxial cable: a cable consisting of two conductors concentric with and insulated from each other.

column: vertical load-carrying member of a structural frame.

commutator: device used on electric motors or generators to maintain a unidirectional current.

concealed: rendered inaccessible by the structure or finish of the building. Wires in concealed raceways are considered concealed, even though they may become accessible by withdrawing them.

conductor: substances that offer little resistance to the flow of electric current. Silver, copper, and aluminum are good conductors although no material is a perfect conductor.

conductor, bare: a conductor having no covering or insulation whatsoever.

conductor, covered: a conductor having one or more layers of non-conducting materials that are not recognized as insulation under the Code.

conductor, insulated: a conductor covered with material recognized as insulation.

connector, pressure (solderless): a connector that establishes the connection between two or more conductors or between one or more conductors and a terminal by means of mechanical pressure and without the use of solder.

continuous load: a load in which the maximum current is expected to continue for 3 hours or more.

contour line: on a land map denoting elevations, a line connecting points with the same elevation.

controller: a device, or group of devices, that serves to govern in some predetermined manner, the electric power delivered to the apparatus to which it is connected.

cooking unit, counter-mounted: an assembly of one or more domestic surface heating elements for cooking purposes, designed to be flush-mounted in, or supported by, a counter and complete with internal wiring and inherent or separately mounted controls.

crawl space: shallow space between the first tier of beams and the ground (no basement).

current: the flow of electricity in a circuit. It is expressed in amperes and represents an amount of electricity.

curtain wall: nonbearing wall between piers or columns for the enclosure of the structure; not supported at each story.

cycle: one complete period of flow of alternating current in both directions. One cycle represents 360°.

demand factor: in any system or part of a system, the ratio of the maximum demand of the system, or part of the system, to the total connected load of the system, or part of the system under consideration.

direct current: current (dc) which flows in one direction only. One wire is always positive, the other negative.

disconnecting means: a device, a group of devices, or other means whereby the conductors of a circuit can be disconnected from their source of supply.

double-strength glass: sheet glass that is ⅛ inch thick (single-strength glass is 1/10 inch thick).

dry wall: interior wall construction consisting of plaster boards, wood paneling, or plywood nailed directly to the studs without application of plaster.

duty, continuous: a requirement of service that demands operation at a substantially constant load for an indefinitely long time.

duty, intermittent: a requirement of service that demands operation for alternate intervals of (1) load and no load, (2) load and rest, or (3) load, no load, and rest.

duty, periodic: a type of intermittent duty in which the load conditions regularly recur.

duty, short-time: a requirement of service that demands operations at loads and for intervals of time, both of which may be subject to wide variation.

efficiency: the name given to ratio of output to input.

electrical generator: a machine so constructed that when its rotor is driven by an engine or other prime mover, a voltage is generated.

electrode: a conducting element used to emit, collect, or control electrons and ions.

electron: a particle of matter negatively charged.

electron emission: the release of electrons from the surface of a material into surrounding space due to heat, light, high voltage, or other causes.

elevation: drawing showing the projection of a building on a vertical plane.

enclosed: surrounded by a case that will prevent anyone from accidentally contacting live parts.

equipment: a general term including material, fittings, devices, appliances, fixtures, apparatus, and the like used as a part of, or in connection with, an electrical installation.

expansion bolt: bolt with a casing arranged to wedge the bolt into a masonry wall to provide an anchorage.

expansion joint: joint between two adjoining concrete members arranged to permit expansion and contraction with changes in temperature.

exposed (as applied to live parts): live parts that a person could inadvertently touch or approach nearer than a safe distance. This term is applied to parts not suitably guarded, isolated, or insulated.

exposed (as applied to wiring method): not concealed.

externally operable: capable of being operated without exposing the operator to contact with live parts.

facade: main front of a building.

farad: a measure of electrical capacity of condensers.

feedback: the process of transferring energy from the output circuit of a device back to its input.

feeder: the conductors between the service equipment, or the generator switchboard of an isolated plant, and the branch circuit overcurrent device.

filter: a combination of circuit elements designed to pass a definite range of frequencies, reducing all others.

fire stop: incombustible filler material used to block interior draft spaces.

fitting: an accessory such as a locknut, bushing, or other part of a wiring system that is intended primarily to perform a mechanical rather than an electrical function.

flashing: strips of sheet metal bent into an angle between the roof and wall to make a watertight joint.

footing: structural unit used to distribute loads to the bearing materials.

frequency: frequency of alternating current is the number of cycles per second. A 60-hertz alternating current makes 60 complete cycles of flow back and forth (120 alternations) per second. A conventional alternator has an even number of field poles arranged in alternate north and south polarities. Current flows in one direction in an ac armature conductor while the conductor is passing a north pole and in the other direction while passing a south pole. The conductor passes two poles during each cycle. A frequency of 60 hertz requires the conductor to pass 120 poles per second. In a 6-pole alternator, the equivalent speed would be 20 revolutions per second or 1200 revolutions per minute. In a 4-pole alternator, the equivalent speed would be 30 revolutions per second or 1800 revolutions per minute.

frostline: deepest level below grade to which frost penetrates in a geographic area.

fuse: a protective device inserted in series with a circuit.

gain: the ratio of output to input power, voltage, or current, respectively.

garage: a building or portion of a building in which one or more self-propelled vehicles carrying volatile, flammable liquid for fuel or power are kept for use, sale, storage, rental, repair, etc.

ground: a conducting connection, whether intentional or accidental, between an electrical circuit or piece of equipment and earth or some other conducting body serving in place of the earth.

grounded: connected to earth or to some conducting body that serves in place of the earth.

grounded conductor: a system or circuit conductor that is intentionally grounded.

grounding conductor: a conductor used to connect equipment or the grounded circuit of a wiring system to a grounding electrode.

henry: the basic unit of inductance.

hertz: a unit of frequency, one cycle per second. Written as 50-hertz or 60-hertz current, etc.

horsepower: the unit of power about equal to the power of a draft horse to do work for a short interval. Numerically, hp is 33,000 ft-lb. per minute; that is, the ability to lift 33,000 pounds to a height of one foot in one minute.

I beam: rolled steel beam or built-up beam of I section.

impedance: effects placed on alternating current by inductive capacitance (current lags voltage), capacitive reactance (current leads voltage), and resistance (opposes current but doesn't lag or lead voltage), or any combination of two. It's measured in ohms like resistance.

incombustible material: material that will not ignite or actively support combustion in a surrounding temperature of 1200°F during an exposure of 5 minutes; also material that will not melt when the temperature of the material is maintained at 900°F for a period of at least 5 minutes.

inductance: the property of a circuit or two neighboring circuits which determines how much voltage will be induced in one circuit by a change of current in either circuit.

inductor: a coil.

integrated circuit: a circuit in which different types of devices such as resistors, capacitors, and transistors are made from a single piece of material and then connected to form a circuit.

isolated: not readily accessible to persons unless special means for access are used.

insulator: substances that offer great resistance to the flow of electric current such as glass, porcelain, paper, cotton, enamel, and paraffin are called insulators because they are practically nonconducting. However, no material is a perfect insulator.

jamb: upright member forming the side of a door or window opening.

kva: the abbreviation of kilovolt-amperes which is the product of the volts times the amperes divided by 1000. This term is used in rating alternating current machinery because with alternating currents, the product of the volts times the amperes usually does not give the true average power.

kvar: the abbreviation of kilovolt-ampere reactance which is a measurement of reactive power that generates power within induction equipment (motors, transformers, holding coils, lighting ballasts, etc.).

kw: the abbreviation for kilowatt which is a unit of measurement of electrical power. A kilowatt (kw) equals 1000 watts and is the product of the volts times the amperes divided by 1000 when used in rating direct current machinery. Also the term used to indicate true power in an ac circuit.

kilowatt-hour: a kilowatt-hour is the amount of electrical power represented by 1000 watts for a period of 1 hour. Thus a generator which delivered 1000 watts for a period of 1 hour would have delivered 1 kilowatt-hour of electricity.

lally column: compression member consisting of a steel pipe filled with concrete under pressure.

laminated wood: wood build up of plies or laminations that have been joined either with glue or with mechanical fasteners. Usually, the plies are too thick to be classified as veneer and the grain of all plies is parallel.

lighting outlet: an outlet intended for the direct connection of a lampholder, lighting fixture, or pendant cord terminating in a lampholder.

location, damp: a location subject to a moderate amount of moisture, such as some basements, some barns, or some cold-storage warehouses.

location, dry: a location not normally subject to dampness or wetness; a location classified as dry may be temporarily subject to dampness or wetness, as in the case of a building under construction.

location, wet: a location subject to saturation with water or other liquids, such as locations exposed to weather or washrooms in garages. Installations that are located underground or in concrete slabs, or masonry in direct contact with the earth shall be considered wet locations.

logic: the arrangement of circuitry designed to accomplish certain objectives.

low energy-power circuit: a circuit that is not a remote-control or signal circuit but whose power supply is limited in accordance with the requirements of Class-2 remote-control circuits.

modulation: the process of varying the amplitude, frequency, or the phase of a carrier wave.

multioutlet assembly: a type of surface or flush raceway designed to hold conductors and attachment plug receptacles and assembled in the field or at the factory.

National Electrical Code: the National Electrical Code is sponsored by the National Fire Protection Association and is the "Bible" of all electrical workers for building construction. It is often referred to as the "NEC" or the "Code."

nonautomatic: used to describe an action requiring personal intervention for its control.

ohm: the unit of measurement of electrical resistance; it represents the amount of resistance that permits current flow at the rate of one ampere under a pressure of one volt. The resistance (in ohms) equals the pressure (in volts) divided by the current (in amperes).

outlet: in the wiring system, a point at which current is taken to supply utilization equipment.

outline lighting: an arrangement of incandescent lamps or gaseous tubes to outline and call attention to certain features such as the shape of a building or the decoration of a window.

oven, wall-mounted: a domestic oven for cooking purposes designed for mounting into or onto a wall or other surface.

panelboard: a single panel or group of panel units designed for assembly in the form of a single panel; includes buses and may come with or without switches and/or automatic overcurrent protective devices for the control of light, heat, or power circuits of small individual as well as aggregate capacity. It is designed to be placed in a cabinet or cutout box placed in or against a wall or partition and accessible only from the front.

pilaster: flat square column attached to a wall and projecting about a fifth of its width from the face of the wall.

plate: the principal anode in an electron tube to which the electron stream is attracted.

plenum: chamber or space forming a part of an air-conditioning system.

potential: the difference in voltage between two points of a circuit. Frequently one is assumed to be ground (zero potential).

potentiometer: an instrument for measuring an unknown voltage or potential difference by balancing it, wholly or in part, by a known potential difference produced by the flow of known currents in a network of circuits of known electrical constants.

power: the rate of doing work or expending energy.

power factor: when the current waves in an alternating-current circuit coincide exactly in time with the voltage waves, the product of volts times amperes gives volt amperes which is true power in watts (or in kw if divided by 1000). When the current waves lag behind the voltage, due to inductive reactance (or lead due to capacitive reactance), they do not reach their respective values at the same time. Under such conditions, the product of volts and amperes does not give true average watts. Such a product is called volt amperes or apparent watts. The factor by which apparent watts must be multiplied to give the true watts is known as the power factor (PF). Power factor depends on the amount of lag or lead, and is the percentage of apparent watts which represents true watts.

Power factor depends on the amount of lag or lead, and is the percentage of apparent watts which represents true watts. With a power factor of 80%, a fully loaded 5-kva alternator will produce 4 kw. When the rating of a power unit is stated in kva at 80% PF, it means that with an 80% PF load, the generator will generate its rated voltage providing the load does not exceed the kva rating.

An engine-driven alternator, for example, with automatic voltage regulation, the kva rating usually is determined by the maximum current which can flow through the windings without injurious overheating or by the ability of the engine or other prime mover to maintain the normal operating speed. A resistance load such as electric lamp bulbs, irons, toasters and similar devices is a unity power factor load. Motors, transformers and various other devices cause a current wave lag which is expressed in the power factor of the load.

precast concrete: concrete units (such as piles or vaults) cast away from the construction site and set in place.

qualified person: one familiar with the construction and operation of the apparatus and the hazards involved.

raceway: any channel designed expressly for holding wire, cables, or bus bars and used solely for this purpose.

raintight: so constructed or protected that exposure to a beating rain will not result in the entrance of water.

reactance: reactance is opposition to the change of current flow in an ac circuit. The rapid reversing of alternating current tends to induce voltages that oppose the flow of current in such a manner that the current waves do not coincide in time with the voltage waves. The opposition of self-inductance to the flow of current is called *inductive reactance* and causes the current to lag behind the voltage which produces it. The opposition of a condensor or of capacitance to the change of alternating current voltage causes the current wave to lead the voltage wave. This is called *capacitive reactance.* The unit of measurement for either inductive reactance or capacitive reactance is the ohm.

readily accessible: capable of being reached quickly, for operation, or inspection, without requiring those to whom ready access is requisite to climb over or remove obstacles or resort to portable ladders, chairs, etc.

receptacle (convenience outlet): a contract device installed at an outlet for the connection of an attachment plug.

receptacle outlet: an outlet where one or more receptacles are installed.

rectifiers: devices used to change alternating current to unidirectional current.

relay: an electromechanical switching device that can be used as a remote control.

remote-control circuit: any electrical circuit that controls any other circuit through a relay or an equivalent device.

resistance: electrical resistance is opposition to the flow of electric current and may be compared to the resistance of a pipe to the flow of water. All substances have some resistance but the amount varies with different substances and with the same substances under different conditions.

resistor: a resistor is a poor conductor used in a circuit to create resistance which limits the amount of current flow. It may be compared to a valve in a water system.

resonance: in a circuit containing both inductance and capacitance, a condition in which the inductive reactance is equal to and cancels out the capacitance reactance.

riser: upright member of stair extending from tread to tread.

roughing in: installation of all concealed electrical wiring; includes all electrical work done before finishing.

saturation: the condition existing in a circuit when an increase in the driving signal doesn't produce any further change in the resultant effect.

sealed (hermetic-type) **motor compressor:** a mechanical compressor consisting of a compressor and a motor, both of which are enclosed in the same sealed housing, with no external shaft or shaft seals, the motor operating in the refrigerant atmosphere.

semiconductor: a material midway between a conductor and an insulator.

service: the conductors and equipment used for delivering energy from the electricity supply system to the wiring system of the premises served.

service cable: the service conductors made up in the form of a cable.

service conductors: the supply conductors that extend from the street main or transformers to the service equipment of the premises being supplied.

service drop: the overhead service conductors from the last pole, or other aerial support, to and including the splices, if any, that connect to the service-entrance conductors at the building or other structure.

service-entrance conductors, underground system: the service conductors between the terminals of the service equipment and the point of connection to the service lateral.

service equipment: the necessary equipment, usually consisting of a circuit breaker, or switch and fuses, and their accessories, located near the point of entrance of supply conductors to a building and intended to constitute the main control and means of cutoff for the supply to that building.

service lateral: the underground service conductors between the street main, including any risers at a pole or other structure or from transformers, and the first point of connection to the service-entrance conductors in a terminal box, meter, or other enclosure with adequate space, inside or outside the building wall. Where there is no terminal box, meter, or other enclosure with adequate space, the point of connection shall be considered to be the point of entrance of the service conductors into the building.

service raceway: the rigid metal conduit, electrical metallic tubing, or other raceway that encloses the service-entrance conductors.

setting (of circuit breaker): the value of the current at which the circuit breaker is set to trip.

sheathing: first covering of boards or paneling nailed to the outside of the wood studs of a frame building.

siding: finishing material that is nailed to the sheathing of a wood frame building and that forms the exposed surface.

signal circuit: any electrical circuit supplying energy to an appliance that gives a recognizable signal.

single phase: a single-phase, alternating-current system has a single voltage in which voltage reversals occur at the same time and are of the same alternating polarity throughout the system.

soffit: underside of a stair, arch, or cornice.

solenoid: an electromagnet having a movable iron core.

soleplate: horizontal bottom member of wood-stud partition.

starter: used only with preheat-type fluorescent lamps to complete a circuit for current flow through the lamp filaments; quickly breaks the circuit when the filaments are sufficiently heated.

studs: vertically set skeleton members of a partition or wall to which lath is nailed.

switch, general-use: a switch intended for use in general distribution and branch circuits. It is rated in amperes and is capable of interrupting its rated voltage.

switch, general-use snap: a form of general-use switch so constructed that it can be installed in flush device boxes or on outlet covers, or otherwise used in conjunction with wiring systems recognized by this Code.

switch, ac general-use snap: a form of general-use snap switch suitable only for use on alternating-current circuits and for controlling the following:

1. Resistive and inductive loads (including electric discharge lamps) not exceeding the ampere rating at the voltage involved.
2. Tungsten-filament lamp loads not exceeding the ampere rating at 120 volts.
3. Motor loads not exceeding 80% of the ampere rating of the switches at the rated voltage.

switch, ac-dc general use snap: A form of general use snap switch suitable for use on either direct or alternating-current circuits and for controlling the following:

1. Resistive loads not exceeding the ampere rating at the voltage involved.
2. Inductive loads not exceeding one-half the ampere rating at the voltage involved, except that switches having a marked horsepower rating are suitable for controlling motors not exceeding the horsepower rating of the switch at the voltage involved.
3. Tungsten-filament lamp loads not exceeding the ampere rating at 125 volts, when marked with the letter T.

switch, isolating: a switch intended for isolating an electric circuit from the source of power. It has no interrupting rating and is intended to be operated only after the circuit has been opened by some other means.

switch, motor-circuit: a switch, rated in horsepower, capable of interrupting the maximum operating overload current of a motor having the same horsepower rating as the switch at the rated voltage.

switchboard: a large single panel, frame, or assembly of panels, having switches, overcurrent and other protective devices, buses, and usually instruments, mounted on the face or back or both. Switchboards are generally accessible from the rear as well as from the front and are not intended to be installed in cabinets.

synchronous: simultaneous in action and in time (in phase).

tachometer: an instrument for measuring revolutions per minute.

thermal cutout: an overcurrent protective device containing a heater element in addition to and affecting, a renewable fusible member which opens the circuit. It is not designed to interrupt short-curcuit currents.

thermal protector (as applied to motors): a protective device that is assembled as an integral part of a motor or motor compressor and that, when properly applied protects the motor against dangerous overheating due to overload and failure to start.

thermally protected (as applied to motors): refers to the words *Thermally Protected* appearing on the nameplate of a motor or motor-compressor and means that the motor is provided with a thermal protector.

three phase: a three phase, alternating-current system has three individual circuits or phase. Each phase is timed so the current alterations of the first phase is ⅓ cycle (120°) ahead of the second and ⅔ cycle (240°) ahead of the third.

transformer: a device used to transfer energy from one circuit to another. It is composed of two or more coils linked by magnetic lines of force.

trusses: framed structural pieces consisting of triangles in a single plane for supporting loads over spans.

utilization equipment: equipment that utilizes electric energy for mechanical, chemical, heating, lighting, or other similar useful purposes.

ventilated: provided with a means to permit circulation of air sufficient to remove an excess of heat fumes, or vapors.

volt: the practical unit of voltage or electromotive force. One volt sends a current of one ampere through a resistance of one ohm.

voltage: the force, pressure, or electromotive force (emf) which causes electric current to flow in an electric circuit. Its unit of measurement is the volt, which represents the amount of electrical pressure that causes current to flow at the rate of one ampere through a resistance of one ohm. Voltage in an electric circuit may be considered as being similar to water pressure in a pipe or water system.

voltage drop: in an electrical circuit, the difference between the voltage at the power source and the voltage at the point at which electricity is to be used. The voltage drop or loss is created by the resistance of the connecting conductors.

voltage to ground: in grounded circuits, the voltage between the given conductor and that point or conductor of the circuit which is grounded; in ungrounded circuits, the greatest voltage between the given conductor and any other conductor of the circuit.

watertight: so constructed that moisture will not enter the enclosing case or housing.

watt: the unit of measurement of electrical power or rate of work. 746 watts is equivalent to 1 horsepower. The watt represents the rate at which power is expended when a pressure on one volt causes current to flow at the rate of one ampere. In a dc circuit or in an ac circuit at unity (100%) power factor, the number of watts equals the pressure (in volts) multiplied by the current (in amperes).

weatherproof: so constructed or protected that exposure to the weather will not interfere with successful operation.

web: central portion of an I beam.

Appendix I

Useful Tables

The following data should prove useful to all electrical contractors. It was prepared by a member of the Minnesota Electrical Association, Inc. for the use of fellow electricians and contractors.

CONDUIT FILL TABLE

The purpose of this table is to find the proper size conduit for combination of conductors not found in Tables 3A, 3B and 3C of the 1975 National Electrical Code. It is based on 3 or more conductors in a conduit and the same tables can be used for either new work or rewiring existing raceways. Types shown are for the most common used, for other insulations, refer to Table 4 and 5 of Chapter 9, 1975 National Electrical Code.

Size Conduit	Unit No.	Wire Size	THHN, THWN	THW	TW
col.1	col.2	col.3	col.4	col.5	col.6
1/2	6	14	.42	1	.66
3/4	10.2	12	.57	1.22	.84
1	16.5	10	.9	1.5	1.1
1¼	29	8	1.6	2.5	2
1½	40	6	2.5	4	4
2	65	4	4.1	5.3	5.3
2½	93.2	3	4.8	6	6
3	143.2	2	5.7	7.2	7.2
3½	192.2	1	7.7	9.8	9.8
4	247	1/0	9.2	11.5	11.5
5	388.4	2/0	11.	13.5	13.5
6	561.2	3/0	13.2	16	16

Example: What size conduit is required for following THW conductors? 3 - #6 and 3 - #12

Column #5

3 - #6 = 3 x 4 = 12.00
3 - #12 = 3 x 1.22 = 3.66
15.66

Column #2 and #1

Total 15.66 is less than 16.5 as found in column 2, so 1" conduit can be used as found in col. 1.

1975 CODE

ALLOWABLE AMPACITIES OF INSULATED CONDUCTORS

Not more than three conductors in raceway or cable or direct burial (base on ambient temperature of 30°C. 86°F.)

SIZE AWG OR MCM	COPPER			ALUMINUM		SIZE AWG OR MCM
	60°C (140°F)	75°C (167°F)	90°C (194°F)	75°C (167°F)	90°C (194°F)	
	TW	THW THWN XHHW USE	THHN XHHW*	THW THWN XHHW USE	THHN XHHW*	
14	15	15	15	–	–	14
12	20	20	20	15	15	12
10	30	30	30	25	25	10
8	40	45	50	40	40	8
6	55	65	70	50	55	6
4	70	● 85	90	65	70	4
3	80	●100	105	75	80	3
2	95	●115	120	◆◆ 90	95	2
1	110	●130	140	◆◆100	110	1
1/0	125	●150	155	◆◆120	125	1/0
2/0	145	●175	185	◆◆135	145	2/0
3/0	165	200	210	◆◆155	165	3/0
4/0	195	230	235	◆◆180	185	4/0
250	215	255	270	205	215	250
300	240	285	300	230	240	300
350	260	310	325	250	260	350
500	320	380	405	310	330	500
600	355	420	455	340	370	600
750	400	475	500	385	405	750
1000	455	545	585	445	480	1000

* For dry locations only, See Table 310-13

For 3-wire, single phase residential services, the allowance ampacity is:

● cu (copper) THW-XHHN-Size 4-100 amp., Size 3-110 amp., Size 2-125 amp., Size 1-150 amp., Size 1/0-175 amp., Size 2/0-200 amp.

◆◆ Al (aluminum) THW-XHHW-Size 2-100 amp., Size 1-110 amp., Size 1/0-125 amp., Size 2/0-150 amp., Size 3/0-175 amp., Size 4/0-200 amp.

Table 370-6(a). Boxes

Box Dimension, Inches Trade Size or Type	Min. Cu. In. Cap.	Maximum Number of Conductors #14	#12	#10	#8	#6
4 x 1¼ Round or Octagonal	12.5	6	5	5	4	0
4 x 1½ Round or Octagonal	15.5	7	6	6	5	0
4 x 2⅛ Round or Octagonal	21.5	10	9	8	7	0
4 x 1¼ Square	18.0	9	8	7	6	0
4 x 1½ Square	21.0	10	9	8	7	0
4 x 2⅛ Square	30.3	15	13	12	10	6*
4 11/16 x 1¼ Square	25.5	12	11	10	8	0
4 11/16 x 1½ Square	29.5	14	13	11	9	0
4 11/16 x 2⅛ Square	42.0	21	18	16	14	6
3 x 2 x 1½ Device	7.5	3	3	3	2	0
3 x 2 x 2 Device	10.0	5	4	4	3	0
3 x 2 x 2¼ Device	10.5	5	4	4	3	0
3 x 2 x 2½ Device	12.5	6	5	5	4	0
3 x 2 x 2¾ Device	14.0	7	6	5	4	0
3 x 2 x 3½ Device	18.0	9	8	7	6	0
4 x 2⅛ x 1½ Device	10.3	5	4	4	3	0
4 x 2⅛ x 1⅞ Device	13.0	6	5	5	4	0
4 x 2⅛ x 2⅛ Device	14.5	7	6	5	4	0
3¾ x 2 x 2½ Masonry Box/gang	14.0	7	6	5	4	0
3¾ x 2 x 3½ Masonry Box/gang	21.0	10	9	8	7	0
FS Minimum Internal Depth 1¾ Single Cover/Gang	13.5	6	6	5	4	0
FD Minimum Internal Depth 2⅜ Single Cover/Gang	18.0	9	8	7	6	3
FS Minimum Internal Depth 1¾ Multiple Cover/Gang	18.0	9	8	7	6	0
FD Minimum Internal Depth 2⅜ Multiple Cover/Gang	24.0	12	10	9	8	4

*Not to be used as a pull box. For termination only.

Table 370-6(b). Volume Required Per Conductor

Size of Conductor	Free Space Within Box for Each Conductor
No. 14	2. cubic inches
No. 12	2.25 cubic inches
No. 10	2.5 cubic inches
No. 8	3. cubic inches
No. 6	5. cubic inches

EVERYDAY INFORMATION

Area of a Circle = Diameter squared x .7854

Area of a Triangle = ½ alt. x base

1 Rod = 5½ yards

Voltage Drop 1ϕ =

$$\frac{D(1\ \text{way}) \times R(\text{see table 8, 70-567, '75 NEC}) \times I}{500}$$

Voltage Drop 3ϕ = V.D. 1ϕ x .866

Converting Temperatures:
(C x 9/5) + 32 = F. (F - 32) x 5/9 = C.

Pulleys — $d = \frac{D \times S}{s}$ or $D = \frac{d \times s}{S}$

d = dia. driven pulley D of driving pulley,
s = RPM of driven S = RPM driving

METRIC

Meter: Approx. 1.1 yard
Liter: Approx. 1.06 liquid quart
Gram: Approx. .04 ounces

COMMON METRIC PREFIXES

Milli: One-thousandth (0.001)
Centi: One-hundredth (0.01)
Kilo: One-thousand times (1000)

EXAMPLES

1000 millimeters = 1 meter
100 centimeters = 1 meter
1000 meters = 1 kilometer

ELECTRIC HEATING WIRING DATA

AMPERE RATINGS, RESISTANCE LOADS
Single Phase

KW	120V.	208V.	240V.	277V.
.50	4.2	2.5	2.1	1.9
.75	6.3	3.7	3.2	2.8
1	8.4	4.9	4.2	3.7
2	16.7	9.7	8.4	7.3
3	25.0	14.5	12.5	10.9
4	33.4	19.3	16.7	14.5
5	41.7	24.1	21.0	18.1
6	50.0	28.9	25.2	21.7
7½	62.5	36.1	31.3	27.1
10	83.4	48.1	41.7	36.2
12	100.0	57.7	50.0	43.4
15	125.0	72.2	62.5	54.2
20	166.7	96.2	83.4	72.3
25	208.4	120.2	104.2	90.3
30	250.0	144.3	125.0	108.4
50	416.7	240.4	208.4	180.6
75	625.0	360.6	312.5	270.8
100	833.4	480.8	416.7	361.1

For kw ratings not listed, combine total of ratings. For example, for 24kw, 240 volts, single phase: 4kw = 16.7 amps; 20kw = 83.4 amp. Total, 24kw = 100.1 amp

AMPERE RATINGS, RESISTANCE LOADS
Three Phase

KW	208V.	240V.	480V.
1	2.8	2.5	1.3
2	5.6	4.9	2.5
3	8.4	7.3	3.7
4	11.2	9.7	4.9
5	13.9	12.1	6.1
6	16.8	14.5	7.3
8	22.4	19.4	9.8
10	27.8	24.1	12.1
12	33.4	29.0	14.5
15	41.7	36.2	18.1
20	55.6	48.2	24.1
25	69.5	60.3	30.2
30	83.4	72.3	36.2
50	139.0	120.5	60.3
75	208.5	180.7	90.4
100	278.0	240.9	120.5

For kw ratings not listed, combine total of ratings. For example, for 9kw, 240 volts, three phase: 4kw = 9.7 amp; 5kw = 12.1 amp. Total, 9kw = 21.8 amp.

ELECTRIC HEAT THUMBNAIL METHOD ESTIMATE

RESIDENTIAL — structures insulated according to Upper Midwest Standards. Following figures based on 20% or less glass or door area.

To Determine (Average) Heat Loss (Single Family)
Watts Loss = 8 watt/ sq. ft. - living areas
Alternate Heat Loss (over 8' ceiling) - 1 watt/cu. ft.
Watts Loss = 5 watt/sq. ft. - Basement

Approximate Cost of Operation
10 x cost/KWHR x sq. ft. = Approx. Yearly Cost

Supplementary Heating — Residential
Watts Loss = 8 watt/sq. ft.

Note: Estimates are based on complete approved insulation. As estimates are **approximate**, don't use for formal proposal. R designates resistance to heat loss — the greater the R value, the better the insulation. Each mfgr. of insulation designates the R value of their product.

Frame Walls: R - 13 - 3-5/8", R - 11 - 3-1/2"
Masonery Walls: R - 10 - 2" styrofoam
Ceiling: R - 36 - Cover Joist 4"
Crawl Space: R - 24 - 6-1/2"
Vapor Barrier: 4 mil-poly (warm side)
Windows: Double Glaze
Sash: Wood/metal Sash with thermal break
Door: Weatherstripped and storm door - 30° outside design

Index